Advanced Sciences and Technologies for Security Applications

The series Advanced Sciences and Technologies for Security Applications comprises interdisciplinary research covering the theory, foundations and domain-specific topics pertaining to security. Publications within the series are peer-reviewed monographs and edited works in the areas of:

- biological and chemical threat recognition and detection (e.g., biosensors, aerosols, forensics)
- crisis and disaster management
- terrorism
- cyber security and secure information systems (e.g., encryption, optical and photonic systems)
- traditional and non-traditional security
- energy, food and resource security
- economic security and securitization (including associated infrastructures)
- transnational crime
- human security and health security
- social, political and psychological aspects of security
- recognition and identification (e.g., optical imaging, biometrics, authentication and verification)
- smart surveillance systems
- applications of theoretical frameworks and methodologies (e.g., grounded theory, complexity, network sciences, modelling and simulation)

Together, the high-quality contributions to this series provide a cross-disciplinary overview of forefront research endeavours aiming to make the world a safer place.

The editors encourage prospective authors to correspond with them in advance of submitting a manuscript. Submission of manuscripts should be made to the Editor-in-Chief or one of the Editors.

More information about this series at http://www.springer.com/series/5540

Laobing Zhang • Genserik Reniers

Game Theory for Managing Security in Chemical Industrial Areas

 Springer

Laobing Zhang
Safety and Security Science Group
Delft University of Technology
Delft, The Netherlands

Genserik Reniers
Safety and Security Science Group
Delft University of Technology
Delft, The Netherlands

ISSN 1613-5113   ISSN 2363-9466   (electronic)
Advanced Sciences and Technologies for Security Applications
ISBN 978-3-030-06473-0   ISBN 978-3-319-92618-6   (eBook)
https://doi.org/10.1007/978-3-319-92618-6

Printed on acid-free paper

This Springer imprint is published by the registered company Springer International Publishing AG part of
Springer Nature.
The registered company address is: Gewerbestrasse 11, 6330 Cham, Switzerland

Introduction

We are convinced that physical security in chemical industrial areas can and should be improved, throughout the world. Chemical substances are stored and processed in large quantities in chemical plants and chemical clusters around the globe, and due to the materials' characteristics such as their flammability, explosiveness, and toxicity, they may cause huge disasters and even societal disruption if deliberately misused. Dealing with security implies dealing with intelligent adversaries and deliberate actions, as will also be further expounded in the next chapters. Such intelligent adversaries require smart solutions and flexible models and recommendations from the defender's side. Such is only possible via mathematical modelling and through the use of game theory as a technique for intelligent strategic decision-making support. In this book, we will elaborate and discuss on how this can be achieved. Figure 1 shows an overview of the book.

Fig. 1 Organization of the book

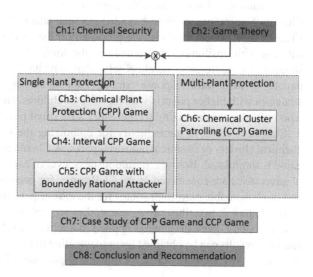

Chapter 1 points out that 'intentionality' is the key difference between a (deliberate) security event and a (coincidental) safety event. The importance of protecting a chemical plant as well as protecting a chemical cluster is illustrated in the chapter. State-of-the-art literature and governmental regulations are discussed. The lack of historical data and the existence of intelligent adversaries are identified as the main challenges for improving security in chemical industrial areas.

Chapter 2 introduces game theory, which is the main methodology used in this book. 'Players', 'strategies', and 'payoffs' are the main components of a game theoretic model. The 'common knowledge' assumption and the 'rationality' assumption are the most frequently used assumptions in game theoretic research and are thoroughly explained. Games with a discrete set of strategies are also discussed (and further used), since they are easier to solve as well as they better reflect reality than games with continuous strategies.

Chapters 3, 4, and 5 concern the physical protection of chemical plants belonging to a single operator. In Chap. 3, a Chemical Plant Protection (CPP) game is developed, based on the so-called multiple-layer protection approach for chemical plants. The CPP game is able to model intelligent interactions between the defender and the attackers. An analysis of the inputs and outputs of the CPP game is also provided.

However, the CPP game suffers a drawback, that is, a large amount of quantitative inputs is required. Chapter 4 therefore addresses this disadvantage, by proposing an Interval CPP game, which is an extension of the CPP game where the exact numbers of the attacker's parameters are no longer needed. Instead, in this game, only the intervals that the parameters will be situated in are required. Thus, the Interval CPP game considers the defender's distribution-free uncertainties on the attackers' parameters, and hence the inputs for the Interval CPP game are easier to obtain, for instance, by using the outputs from the API SRA method [1].

A second drawback of the CPP game concerns the rational attacker assumption. Chapter 5 therefore models bounded-rational attackers into the CPP game. In Chap. 5, three robust solutions are proposed for the CPP game, namely, the Robust solution with epsilon-optimal attackers, the MoSICP solution, and the MiniMax solution, for addressing attackers who may deviate from strategies having close payoffs to their 'best response' strategy, for addressing attackers who may play strategies with higher payoffs with higher probabilities, and for addressing attackers who only aim at minimizing the defender's maximal payoffs, respectively.

Chapter 6 employs game theory for optimizing the scheduling of patrolling in chemical clusters or chemical industrial parks. A Chemical Cluster Patrolling (CCP) game is formulated. Both the hazardousness level of each plant and the intelligence of adversaries are considered in the CCP game, for generating random but strategic and implementable patrolling routes for the cluster patrolling team.

In Chapter 7, two illustrative case studies are elaborated and investigated. In the first case study, the CPP game is applied to a refinery to show how the game works and what results can be obtained by implementing the game. The refinery case is also used in the API SRA document for illustrative purposes. Therefore, the outputs from the API SRA method are used as one part of the inputs for the CPP game, while other inputs of the CPP game are illustrative numbers. In the second case study, the CCP game is applied to a chemical cluster composed of several plants, each belonging to different operators, for optimizing the patrolling of security guards in the multi-plant area. Results show that the patrolling route generated by the CCP game well outperforms the purely randomized patrolling strategy as well as all the fixed patrolling routes.

Eight conclusions are drawn and nine recommendations are given in Chap. 8.

Reference

1. API. Security risk assessment methodology for the petroleum and petrochemical industries. In: 780 ARP, editor. 2013.

Contents

List of Figures

Chapter 1
Protecting Process Industries from Intentional Attacks: The State of the Art

1.1 Introduction

Large inventories of hazardous chemicals which can cause catastrophic consequences if released maliciously, the presence of chemical agents which can be stolen and be used either in later terrorist attacks or in making chemical and biochemical weapons, along with the key role of chemical plants in the economy and the public welfare and as an integral element in the supply chain have made the security of chemical plants a great concern especially since 9/11 terrorist attacks in the US. Aside from the importance of chemical plants themselves as potentially attractive targets to terrorist attacks, the usage of chemicals in more than half of the terrorist attacks worldwide further emphasizes the security assessment and management of chemical plants.

The terrorist attacks to chemical facilities (excluding the ones located in war zones) have been very few and far between (Table 1.1 [1]). Yet, the risk of terrorist attacks should not be underestimated by authorities and plants' owners and security management; attacks to two chemical facilities in France in June and July 2015 raised a red flag about the imminent risk of terrorist attacks to chemical plants in the Western world.

Aside from the regulations, standards, and guidelines set forth by, among others, the Centre of Chemical Process Safety (CCPS) of the American Institute of Chemical Engineers in 2003 ("Guidelines for Analyzing and Managing the Security Vulnerabilities of Fixed Chemical Sites"), American Petroleum Institute (API) in 2003 and renewed in 2013 (Security Vulnerability Assessment Methodology for the Petroleum and Petrochemical Industries), and The Chemical Facility Anti-Terrorism Standards (CFATS) in 2007 and renewed in 2014, still many chemical facilities in the US containing Chemicals of Interest (COI) as denoted in the Appendix A of CFATS are not willing to submit a Top Screen consequence assessment to the US Department of Homeland Security (DHS). Not to mention that the lack of relevant

© Springer International Publishing AG, part of Springer Nature 2018
L. Zhang, G. Reniers, *Game Theory for Managing Security in Chemical Industrial Areas*, Advanced Sciences and Technologies for Security Applications,
https://doi.org/10.1007/978-3-319-92618-6_1

Table 1.1 Terrorist attacks to chemical facilities

Year	Country	Target
1974	Greece	DOW Chemicals
1983	Peru	Bayer Chemicals
1984	India	Pesticide plant
1985	Belgium	Bayer Chemicals
1990	Libya	The Rabta Chemicals
1997	US	Natural-gas processing facility
2000	US	Propane storage facility
2001	Yemen	Nexen chemicals company
2002	Colombia	Protoquimicos Company
2003	Russia	Storage tanks
2005	Iraq	Natural gas pipelines
2005	Turkey	Polin Polyester factory
2005	Spain	Paint factory
2013	Algeria	Tigantourine gas facility
2005	Spain	Metal works facility
2015	France	Chemicals company
2015	France	Storage tanks
2016	Algeria	Krechba gas facility
2016	Iraq	Taji gas plant and energy facility
2016	Iraq	Chemical plant
2016	Lybia	Oil storage tank facilities in El-sider

regulations and unwillingness of the chemical and process industries in European countries and in the developing countries to establish and implement security risk assessment and management, is much more severe.

1.2 Safety and Security Definitions and Differences

Definition

Safety and security are two related concepts but they have a different basis. Table 1.2 gives an overview of various definitions for safety and security. A distinction is made between definitions that focus on specific properties and definitions that focus on global properties.

Safety and security are different in the nature of incidents: safety is non-intentional, whereas security is intentional (and related with deliberate acts). This implies that in the case of security an aggressor is present who is influenced by the physical environment and by personal factors. These parameters should thus be taken into account during security assessments. The aggressor may act from within the organization (internal) and from outside the organization. Probabilities in terms of security are very hard to determine. Hence, the identification of threats and the development of measures in terms of security is a challenging task.

Table 1.2 Definitions of safety and security from a specific and a global viewpoint

Safety	Specific properties	Protection against human and technical failure
		Harm to people caused by arbitrary or non-intentional events, natural disasters, human error or system or process errors
	Global properties	Protecting the environment
Security	Specific properties	Protect against deliberate acts of people
		Loss caused by intentional acts of people
		Intentional human actions
	Global properties	Prevent a disruption of services and critical sectors
		Securing the whole environment

Table 1.3 Non-exhaustive list of differences between safety and security

Safety	Security
The nature of an incident is an inherent risk	The nature of an incident is caused by a human act
Non-intentional	Intentional
No human aggressor	Human aggressor
Quantitative probabilities and frequencies of safety-related risks are often available	Only qualitative (expert-opinion based) likelihood of security-related risks may be available
Risks are of rational nature	Threats may be of symbolic nature

Both concepts also differ in their approach. In case of safety assessments (or so-called 'risk analyses'), risks are detected and analyzed by using consequences and probabilities (or frequencies). In case of security risk assessments (or so-called 'threat assessments'), threats are detected and analyzed by using consequences, vulnerabilities and target attractiveness. The different approach sometimes leads to the need for different and complementary protection measures in case of safety and security. Table 1.3 provides an overview of different characteristics attached to safety and to security.

In summary, while safety risks concern possible losses caused by non-intentional events, such as natural disasters, failure of aging facilities, and mis-operations, etc., security risks are related to possible losses caused by intentional human behaviour, such as terrorist attacks, sabotage by disgruntled employees, criminals, etc.

The Importance of the Differences Between Safety and Security

A key difference, amongst others, between safety risks and security risks is whether there are intelligent interactions between the risk holder and the risk maker. "Intelligent interactions", in this statement, means that the risk maker must have the ability to schedule his behaviour to meet his own interests, according to the risk holder's behaviour. In a safety event, due to the mere characteristics of such event as explained in the previous section, risk makers do not have the ability to plan their behaviour.

For instance, a typical type of safety event is a natural disaster, such as an earthquake, a flood, extreme weather etc. In this kind of events, nature can be seen as the risk maker. The risk holders are targets (for instance, people, property, reputation, etc.) who suffer losses from these events. The risk holder may defend itself against nature (e.g., build higher dams or use lightning deflectors), but the risk maker, nature in our example, does not have its own interests and hence does not plan its behaviour.

A more complicated example is that the risk initiator behaves in a way that he would like to achieve a goal, but non-intentionally causes an unplanned accident. A typical scenario of this situation can be a thief stealing a computer from an organization for obtaining the hardware device, and accidently he steals a computer with important technical and confidential information (without backup available). This scenario concerns a security risk since it satisfies the following conditions: (i) the thief has the ability to plan his behaviour according to the organization's defence; and (ii) the thief has his own interests to meet.

The most difficult part of distinguishing a safety event from a security event is to judge whether the risk maker has his own interests with respect to the event or not. An industrial accident caused by a mis-operation, for example, is defined as a safety event. Nevertheless, an accident caused by a disgruntled employee (thus causing intentional mis-operation) would be defined as a security event. In both events, the risk maker has the ability to plan his action. However, in case of the coincidental mis-operation (without the aim to cause losses), the employee does not have his own interest in causing the event and doesn't obtain anything from the event. In case of the disgruntled employee, the employee's interest is to obtain mental satisfaction from the event. This theoretical difference makes it extremely difficult in some cases to distinguish whether an accident can be classified as a security event or as a safety event.

The risk maker from a security viewpoint, although being able to behave according to the risk holder's behaviour, doesn't necessarily do so, and thus doesn't need to act intelligently. To have the ability to act intelligently is one thing, while to use this ability is another thing. Therefore, in security events, we may also see some random behaviour. For instance, an attacker with so-called 'bounded rationality' does exist in the real world. Furthermore, whether the risk maker (actually) behaves randomly is not a clear criteria to unambiguously decide whether the event can be classified as a safety or as a security related event. As an obvious example of this reasoning, in a terrorist attack scenario, when the defender enhances her defence, the attacker is supposed not to implement an attack any more. However, the attacker can behave irrationally (see also definition of 'rationality' in Sect. 2.1.6), and despite the extra defence measures, attack the defender anyway.

1.3 Security in a Single Chemical Plant

1.3.1 The Need of Improving Security in Chemical Plants

Security research has a long history. It has obviously been stimulated by the 9/11 attack in New York in 2001, and ever since, people ever more perceive terrorism as an urgent problem. Figure 1.1 illustrates the yearly number of global terrorist attacks (Source: Global Terrorism Database [2]). Hence, despite a number of academic studies and societal financial efforts for preventing terrorist attacks, the figure shows that the global amount of terrorist attacks sharply increased during the past decade.

Moreover, our highly connected modern Western societies are vulnerable and fragile to possible targeted attacks. Many networked sub-systems of the modern society such as the internet, interlinked financial institutions, airline networks, etc., satisfy the so-called "power-law" degree distribution. This means that only few nodes in these networks exhibit a high degree of importance in the network if compared to most other nodes belonging to the network. If these high-importance nodes would be intentionally attacked, the network would suffer severely.

In the process industries, we see that on the one hand chemical plants tend to 'cluster' together in industrial parks and to build geographically close to each other, due to all kinds of benefits of scale. However, due to the existence of so-called 'domino effects' [3] if one plant or installation would be attacked intelligently, the whole cluster as well as its surrounding area could be affected. On the other hand, plants/companies are also highly dependent on their upstream and downstream plants, through the supply chain. Thus if one plant would be attacked and stops its operation, many more plants would be economically affected as well.

Summarizing the above observations, not only the frequency of terrorist attacks seems to be increasing, but due to the characteristics of our modern societies and the interconnectiveness between people and between companies, also the potential devastation of malicious attacks is growing.

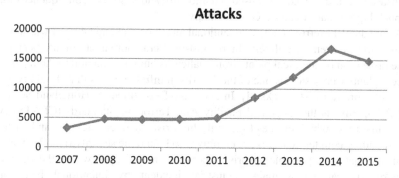

Fig. 1.1 The trend of global terrorist attacks from 2007 to 2015

Chemical and process plants have important roles for our modern way of life. They provide materials for our clothes, food, medicines etc. Chemical industries also form the foundation of modern transportation systems, by providing energies (mainly oil and gas) and stronger materials. Moreover, considering the fact that the chemical industry can be seen as the foundation of a lot of other industries, e.g., the manufacturing industry, its role in the regional economic surrounding cannot be underestimated.

Besides its importance for our modern way of live, the chemical industry may also pose an important threat to today's society. Toxic and flammable materials, as well as extreme pressure and temperature conditions, may be involved in production processes. Therefore, if these materials are not operated and managed correctly, and/or the extreme production conditions are not controlled well, disastrous events might result. Many disasters can be mentioned as examples. For instance, Seveso in 1976 and Bhopal in 1984 are examples of the leakage of toxic gas causing huge consequences for industry and society. The Mexico City disaster in 1984 is an example of the worst ever happened domino effect, causing 650 casualties [3]. Other true disasters causing detriment and devastation include Flixborough in 1974, Basel in 1986, Piper Alpha in 1987, Nagothane in 1990, Toulouse in 2001, Texas City in 2005, Buncefield in 2005, Deepwater Horizon in 2010, etc.

All these abovementioned disasters were initiated by coincidence (for example, misoperation or poor industrial management), and therefore they can be classified as safety events. If intentional attacks would have been involved in these disasters, they would have been even more difficult predictable and their consequences could in most cases be even higher. Actually, the worst ever industrial accident that happened in the chemical industry is the Bhopal gas tragedy in 1984, and the company operating the Bhopal plant at that time has always claimed that this disaster was a security event. However, the accident has been extremely thoroughly investigated, and we now know without any doubt that it was a safety related event. Nonetheless, two important observations can be made from this example: (i) the fact that the company always claimed that the event was security related indicates that without thorough investigation it is difficult to be sure of the nature of a disaster, and (ii) disasters could indeed be caused intentionally and if so, the consequences may be much higher than if caused coincidentally.

Before the 9/11 terrorist act, an intentional attack on a chemical plant was always believed to be extremely unlikely. In the post-9/11 era, more attention has been paid to the protection of chemical plants from malicious human behaviour. Chemical and process plants were listed as one of the 16 critical infrastructures that should be well protected from terrorist attacks [4]. In 2007, the Department of Homeland Security (DHS) implements the Chemical Facility Anti-Terrorism Standards (CFATS) Act for the first time, which obliges to identify high-risk chemical facilities and ensures corresponding countermeasures are employed to bound the security risk. Pasman [1] points out that three possible terrorism operations may happen within the chemical industry: (i) causing a major industrial incident by intentional behaviour, for example, by using a bomb or even simply by switching off a valve; (ii) disrupting the production chain of some important products, e.g., medicines; and (iii) stealing materials for a further step attack, e.g., obtaining toxic materials and release it in a public place.

Anastas and Hammond [5] indicate that across the United States, approximately 15,000 chemical plants, manufacturers, water utilities, and other facilities store and use extremely hazardous substances that would injure or kill employees and residents in nearby communities if suddenly released. Approximately 125 of these facilities each put at least 1 million people at risk; 700 facilities each put 100,000 people at risk; and 3000 facilities each put at least 10,000 people at risk, cumulatively placing the well-being of more than 200 million American people at risk. Hence, the threat of terrorism has brought new scrutiny to the potential for terrorists to deliberately trigger accidents that until recently the chemical industry characterized as extremely unlikely worst-case scenarios. Nevertheless, a single terrorist attack could have even more severe consequences than the thousands of accidental releases that occur and the many people that suffer each year as a non-intended by-product of ongoing use of hazardous chemicals. A large-scale European study in this regard has not yet been carried out, but the figures and numbers of risk makers (chemical plants) and risk holders (potential victims) in Europe are most likely similar, or even higher, to those of the United States. In Europe, approximately 12,000 chemical plants are situated.

In Iraq, frequent attacks to oil pipelines and refineries caused more than 10 billion dollars in the period 2003–2005 [6]. Furthermore, an analysis carried out by Khakzad [1] reveals that chemicals are involved in more than half of the terrorist attacks which happened in the world in 2015.

Reniers and Pavlova [7] categorize accidents into three different types, namely Type I, Type II and Type III, according to the available historical data of these accidents. Type I accidents are accidents with abundant data, and are mainly referring to individual level events, such as falling, slipping, little fires etc. Type II accidents are accidents with extremely/very little records of data, and are mainly referring to industrial disasters, such as the Bhopal disaster, the Seveso disaster etc. Type III accidents are accidents with no historical data at all, so-called black swans, and are mainly referring to accidents where multiple plants are involved. Type III accidents can however be seen as the extremum of Type II accidents. In security terminology, Type I events can be seen as thefts, manslaughter and murder, while Type II events are terrorist attacks.

Reniers and Khakzad [8] further argue that although two safety revolutions happened in the last century, dramatically reducing the number of Type I accidents, a new revolution is needed for further reducing the Type II accidents. Moreover, previous methodologies and theories for reducing Type II events are mainly conducted from a safety point of view. In the post-9/11 era, accidents initiated by intentional behaviour should also be considered, and if so, one can no longer be confident to say that the probability of a Type II event is extremely low.

1.3.2   Challenges with Respect to Improving Chemical Security

Two challenges make security research in chemical plants particularly difficult: (i) the lack of research data (statistical historic data or experimental data); and (ii) the existence of intelligent adversaries.

Security events, in particular terrorist attack events, do not happen frequently in chemical plants, and for those that did happen, the data collection is not sufficient. Therefore, only scarce security data is available. To make matters even more difficult, most security related data is protected very well, at least to the public and to academic researchers. Due to the lack of available data, statistical models and methods for modelling risk makers' behaviour are not applicable. Statistical modelling has nonetheless a long history of being used in the safety domain. For instance, by collecting data, industrial managers know which segment of a pipeline is the most vulnerable part.

Statistical modelling may also be used in the security domain. For instance, by collecting the number of detected intruders, we can evaluate the efficiency of the intrusion detection system (IDS). In any case, statistics-based learning doesn't work when there are only a limited number of records. Furthermore, intruders might be deterred due to an enhanced IDS, which will further reduce the number of detected intruders.

The existence of intelligent adversaries is another challenge for improving security. As we stated in the previous section, security risk makers would plan their behaviour according to the risk holder's defence, in order to meet the risk maker's own interests. Therefore, in security events, the defender has to always take the attacker's response into consideration. Figure 1.2 illustrates how resources can be

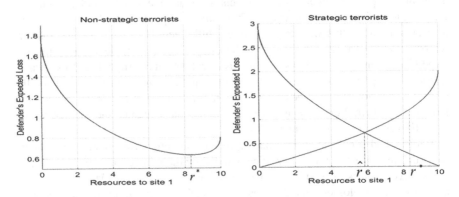

Fig. 1.2 Security investment w.r.t. strategic vs. nonstrategic terrorist

mis-allocated if the defender does not take intelligent attackers into account. In Fig. 1.2, comparison of security investments to a non-strategic terrorist (the left hand side figure) and to a strategic terrorist (the right hand side figure) is shown. Ten resources are being allocated to two sites which values three and two respectively. The curve in the left hand figure is plotted as $DEL = \alpha_1 \cdot L_1 \cdot v_1(r) + \alpha_2 \cdot L_2 \cdot v_2(R - r)$, which denotes the SVA methodology. The curves in the right hand side figure are plotted as $DEL1 = L_1 \cdot v_1(r)$ and $DEL2 = L_2 \cdot v_2(R - r)$, for the decreasing curve and for the increasing curve respectively, and they denote the game theoretic results. It reveals that the SVA methodology without considering the strategic terrorists suggests to allocate $r^* \approx 8.3$ resources to site 1 while the game theoretic model which models the intelligent interactions between the defender and the attacker, suggests to allocate $\hat{r} \approx 5.8$ resources to site 1. This figure was adopted from Powell [9].

Moreover, the existence of intelligent adversaries also highlights the challenge with respect to the lack of data. Since security adversaries are so-called 'intelligent', the statistical data based approach, if being used in security risk assessment, can be misleading. For instance, some security risk assessment methods also try to employ a data based approach for predicting security events. The API SRA standard [10], among others, suggests a historic data based approach for estimating threat ranking for the chemical industries. According to the API SRA standard, most chemical plants have the same – very low – level of terrorist threat ranking, since most of them have "no expected attack in the life of the facility's operation". However, whether an intelligent attacker would attack the plant or not, does not depend much on the historic data, instead, it depends on whether the plant can meet their own interest and on whether their attack on the plant would easily be successful or not.

Furthermore, it is difficult to collect experimental data for behaviour modelling of an intelligent adversary. Security adversaries would not join any security experiments and they can hide their behaviours during the experiments as well. For instance, for a safety research purpose, psychological experiments can be employed to estimate the probability of human errors in different situations. However, if this experiment would be carried out for a security purpose, then finding attacker participants is difficult (if not impossible) and if ordinary people would be invited to act as attackers, the data would not be reliable since attackers and ordinary people can behave totally differently.

1.3.3 Security Risk Assessment in Chemical Plants: State-of-the-Art Research

The risks of deliberate acts to cause losses are addressed using security risk assessment (SRA) to determine if existing security measures and process safe guards are adequate or need improvement [11]. Conceptually, a security risk can be viewed as the intersection of events where threat, vulnerability and consequences are present. This can be compared with a safety event which can be regarded as the triangle of

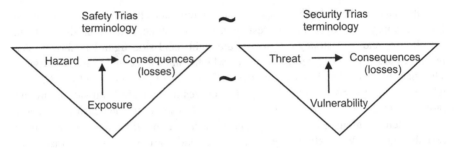

Fig. 1.3 Safety trias and security trias

hazard, exposure and consequences [12]. Figure 1.3 illustrates this conceptualization and comparison of safety and security risks.

Risk assessment consists of hazard identification, risk analysis, and risk evaluation. Hazard identification involves the identification of risk sources, events, their causes and potential consequences. Risk analysis is used to determine the level of risk, using a pre-determined qualitative or quantitative calculation method. Risk evaluation is the process of comparing the results of risk analysis with certain risk criteria to determine whether the risk is tolerable or acceptable, or not. It assists in the decision about risk treatment to reduce risk, if needed.

Hazard identification is the starting point for risk assessment. It equates to process hazard analysis PHA in the safety domain [13] and security vulnerability analysis (SVA) in the security domain [14–19]. Baybutt [1] indicates that SVA is the security equivalent of PHA. It involves evaluating threat events and/or threat scenarios. They originate with hostile action to gain access to processes in order to cause harm. A threat event pairs an attacker and their intent with the object of the attack. A threat scenario is a specific sequence of events with an undesirable consequence resulting from the realization of a threat. It is the security equivalent of a hazard scenario. Generally, a threat event represents a set of threat scenarios. and security risk assessment depends on the completeness of scenario identification in SVA. If scenarios are missed, security risks will be underestimated.

Baybutt [1] recommends that prior to performing SVAs, companies should take remedial measures to protect their facilities that are obvious without the need to conduct an SVA, for example, for physical security: inventory control, personnel screening, security awareness, information control, physical barriers, surveillance systems, and access controls; and for cyber security: personnel screening, firewalling control systems, air gapping safety instrumented systems, eliminating or controlling/securing modems, managing portable computer storage media, etc. Such issues can be addressed by facility audits before SVAs are performed.

SVA usually addresses high-risk events with potentially catastrophic consequences such as those that may arise as a result of terrorist attacks. Typically, these involve large-scale impacts that could affect a significant number of people, the public, the facility, the company, the environment, the economy, or the country's infrastructure (industrial sectors needed for the operation of the economy and government). However, SVA also can be used to address other plant security risks such as the theft of valuable process information for financial gain.

An SVA for a facility endeavors to address these questions [20]:

- Will a facility be targeted?
- What assets may be targeted?
- How may assets be exploited?
- Who will attack?
- How will they attack?
- What protection is there against an attack?
- What will be the consequences?
- Is additional protection needed?

The overall objectives of SVA are to identify credible threats to a facility, identify vulnerabilities that exist, and provide information to facilitate decisions on any needed corrective actions that should be taken. SVA uses structured brainstorming by a team of qualified and experienced people, a technique that has a long history of success in the safety field. It has been noted that identifying scenarios for risk analysis is part science and part art [21]. SVA requires the application of creative thinking [22] to help ensure the completeness of threat and vulnerability identification and critical thinking [23, 24] to help ensure that the results are not subject to cognitive or motivational biases [25, 26]. The underlying model for the analysis is depicted in Fig. 1.4 (Source: Baybutt, 2017 [20]).

A variety of SVA methods have been developed to identify and analyze threats and vulnerabilities of process plants to attacks. They share a number of points and they all address assets to be protected. They differ only in the approach taken.

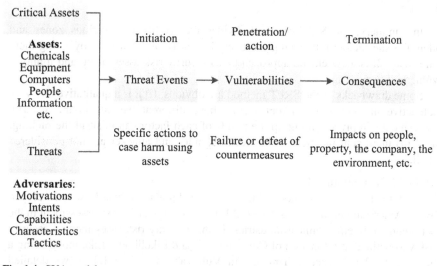

Fig. 1.4 SVA model

Historically, two philosophically different SVA approaches were developed for physical security: asset-based and scenario-based. The asset-based approach originated with security professionals who focus efforts on protecting valuable assets. The scenario-based approach originated with safety professionals who focus on protecting against accidents and the scenarios they involve. Both approaches consider how assets can be exploited by adversaries to cause harm [1].

SVA methods are performance-based and do not require the use of any specific risk remediation measures or countermeasures. SVA studies must be documented to allow review by peers and others. Often SVA study results are recorded in the form of a spreadsheet which offers the benefit of easy updating when needed. The format of the analyses is similar to PHA and, therefore, the methods offer the further benefit of familiarity to individuals who have participated in PHAs, a number of whom will likely also be members of SVA teams.

When looking at chemical security, different SVA methods are possible. Amongst others, two systematic methods can be mentioned: the Security Risk Factor Table (SRFT) [16] and the Security Vulnerability Assessment Methodology (SVA) [10].

SRFT
In 2002, SRFT was first proposed by the "Advanced Chemical Safety Company" to carry out a security risk assessment for a given chemical facility. The basic idea is to identify security-related factors of the given facility, to rate them on a scale from 0 to 5, with 0 being the "lowest risk" and 5 being the "highest risk", and finally to sum up the scores of each factor to measure the security risk status of the facility. Figure 1.5a shows an example of a part of an SRFT table [16]. In the example given by Bajpai, the chosen factors are Location/Visibility/Inventory etc.; for each factor, scoring criteria are given, and each factor obtains a score of 1, 2, 5 etc.; the total score of this facility is 35. He finally concludes that security risk of this plant is High, according to Fig. 1.5b.

In summary, the SRFT method divides the facility into various zones and identifies the factors influencing the overall security of the facility by rating them on a scale. It is a systematic approach to do security risk assessment, and it allows vulnerability ranking.

Some drawbacks of the SRFT method are obvious: (i) it is a qualitative and very subjective method; (ii) different factors have different weights in the security assessment, simply summing up the points of each factor can mislead the ranking; and (iii) intelligent interactions between defender and attacker are not considered at all.

The API SRA Standard
In 2003, the first "SVA method" as it has become known afterwards, was developed by the American Petroleum Institute (API) to perform security risk assessment in the petroleum and petrochemical industries. In this security risk assessment, a security risk was defined as a function of Consequences and Likelihood; Likelihood being a function of Attractiveness, Threat, and Vulnerability. Table 1.4 shows detailed

Risk Factors	Range of Security Points			Points
Location	Rural 1	Urban 2,3,4	High density 5	1
Visibility	Not visible 0	Low Medium 1,2 3,4	High 5	2
Inventory	Low 1	Medium Large 2 3,4	Very Large 5	5
Ownership	Private 1	Public/Co-operative 2,3	Government 4,5	3
Presence of chemicals which can be used as precursor for WMD	Absence 0		Presence 5	0

(a) SRFT

CSRS *	Actual Points Obtained	Recommendations
Low	<15	Maintain security awareness without excessive concern.
Moderate	16-30	Review and update existing security procedures in light of possible threats.
High	31-45	Identify risk-drivers that can be reduced with reasonable controls. Conduct threat & vulnerability analysis and work with law enforcement agencies to enhance security.
Extreme	>45	Initiate aggressive risk-reduction activity, in conjunction and consultation with law enforcement agencies. Conduct threat and vulnerability analysis.

(b) Security Risk Ranking

Fig. 1.5 SRFT example from Bajpai (CSRS: Current Security Risk Status)

Table 1.4 Definitions of terminologies in the API SRA method

Terminology	Definition
Consequence	The degree of injury or damage that would result if there were a successful attack
Threat	Any indication, circumstance or event with the potential to cause loss of, or damage, to an asset
Vulnerability	Any weakness that can be exploited by an adversary to gain unauthorized access and subsequent destruction or theft of an asset
Attractiveness	An estimate of the real or perceived value of a target to an adversary

definitions of some terminologies used in this first so-called SVA method. The SVA methodology consists of 5 steps: (i) Characterization- Characterize the facility or operation to understand what critical assets need to be secured, their importance, their infrastructure dependencies and interdependencies; (ii) Threat Assessment- Identify and characterize threats against those assets, and evaluate the assets in terms of attractiveness of the targets to each threat and the consequences if they are damaged, compromised, or stolen; (iii) Vulnerability Assessment- Identify potential security vulnerabilities that enhance the probability that the threat would successfully accomplish the act; (iv) Risk Assessment- Determine the risk represented by these events or conditions by determining the likelihood of a successful event and the maximum credible consequences of an event if it were to occur; rank the risk of the event occurring and, if it is determined to exceed risk guidelines, make the recommendations to risk reductions; (v) Countermeasures analysis- Identify and evaluate risk mitigation options (both net risk reduction and benefit/cost analyses) and re-assess risks to ensure the adequate countermeasures are being applied. Evaluate the appropriate response capabilities for security events and the ability of the operation or facility to adjust its operations to meet its goals in recovering from the incident. In 2013, API published a new version of SVA and in this version, SVA was named as Security Risk Assessment (SRA). But the basic terms and steps are the same. Hereafter in the book, we name this methodology as "the API SRA methodology".

Figure 1.6 in combination with Table 1.5, briefly illustrate the security risk assessment and management procedure of the API SRA methodology. The left-hand side of Fig. 1.6 shows the sub-steps of the methodology, while the right-hand side shows the output data of each step. Explanations of the outputs are given in Table 1.5.

In the characterization step, the SRA team roughly scans the given petrochemical plant, and provides a critical assets list CAL as well as asset severity scores AS, according to functions of assets, interconnectivities among assets, and possible consequences. In the threat assessment step, the SRA team decides a threats list TL and threat levels TS that the plant is faced with, based on historical security data (site-specific, national, worldwide) and intelligence. For each asset and threat pair $\{(a, t)| a \in CAL, t \in TL\}$, the asset's attractiveness to the threat $Atr_{(a, t)}$ and possible attack scenarios linking the threat with the asset $Sce_{(a, t)}$ are evaluated. Based on current (situation '1') security countermeasures, vulnerabilities $V^1_{(a,t,s)}$ and consequences $C^1_{(a,t,s)}$ are estimated for each asset, threat, and scenario triad $\{(a, t, s)|$ $a \in CAL, t \in TL, s \in Sce_{(a, t)}\}$. Furthermore, the SRA team calculates the likelihood of an attack from a given threat $t \in TL$ to a given asset $a \in CAL$ as $L^1_{(a,t)} = TS_t \times Atr_{(a,t)}$, and calculates the likelihood of a successful attack from t to a by using scenario $s \in Sce_{(a, t)}$ as $L_{(a,t,s)} = L^1_{(a,t)} \times V^1_{(a,t,s)}$. The risk matrix method is used to calculate a security risk $R^1_{(a,t,s)}$ for each asset, threat, and scenario triad, and in this step, the likelihood of a successful attack $L_{(a, t, s)}$ and the scenario-specific consequence $C^1_{(a,t,s)}$ are used to determine the risk value in the risk matrix. Based on the gaps between the current security risk and the desirable level of risk, scenario-specific countermeasures $CM_{(a, t, s)}$ are proposed by the SRA team, and subsequently all the scenario-specific countermeasures are united into one countermeasure list CML.

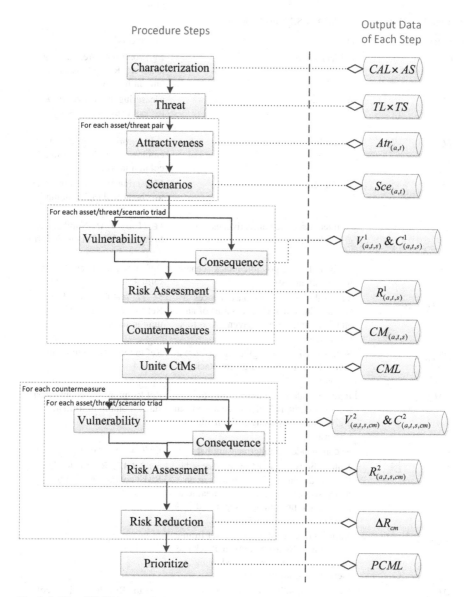

Fig. 1.6 The API SRA procedure

The SRA team further re-estimates the vulnerabilities $V^2_{(a,t,s,cm)}$, consequences $C^2_{(a,t,s,cm)}$, and security risks $R^2_{(a,t,s,cm)}$, presuming that a countermeasure $cm \in CML$ is implemented (situation '2'). Based on the recalculation, the risk reduction of each countermeasure ΔR_{cm} can be calculated as the summation of risk reduced in each asset, threat, and scenario triad, as shown in Formula (1.1). Finally, the proposed

Table 1.5 Output data of the API SRA methodology

Notation	Definition	Comments[a]
CAL	Critical assets list	e.g., control centre, gasoline tanks etc. Ref to "assets" column in form 1.
AS	Asset score	Measuring asset severity. Ref to "asset severity ranking" column in form 1.
TL	Threat list	e.g., terrorists, disgruntled employee etc. Ref to "threat" column in form 2.
TS	Threat score	Measuring threat ranking. Ref to "threat ranking" column in form 2.
$At_{(t,a)}$	A given asset's (a) attractiveness to a given threat (t).	$t \in TL$, $a \in CAL$. Numbers, ref. to column 2a1, 2b1 etc. in form 3.
$Sce_{(a,t)}$	A given threat's possible attack scenarios to a given asset.	Ref to "scenario" column in form 4.
$V^1_{(a,t,s)}$, $C^1_{(a,t,s)}$	Vulnerability '1' and Consequences '1' (in case the attack is successful) of an attack scenario from a given threat to a given asset.	$t \in TL$, $a \in CAL$, $s \in Sce_{(a,t)}$. Ref to the "V" and "C1" column in form 4.
$R^1_{(a,t,s)}$	Security risk '1' of a given asset from a given threat by using a given attack scenario.	Ref to the "R1" column in form 4.
$CM_{(a,t,s)}$	Recommended countermeasures to reduce security risk of a given asset from a given threat by using a given attack scenario.	Ref to "proposed countermeasures" column in form 4.
CML	Recommended countermeasure list	$CML = \bigcup_{a,t,s} CM_{(a,t,s)}$.
$V^2_{(a,t,s,cm)}$, $C^2_{(a,t,s,cm)}$	Vulnerability '2' and Consequences '2' (in case the attack is successful) of an attack scenario from a given threat to a given asset, presuming a suggested countermeasure is implemented.	$cm \in CML$. Ref to "Residual Risk" column in form 5.
$R^2_{(a,t,s,cm)}$	Security risk '2' of a given asset from a given threat by using a given attack scenario, presuming a suggested countermeasure is implemented.	
ΔR_{cm}	Risk reduction by a proposed countermeasure.	Ref to "Risk Reduction" column in form 6.
$PCML$	Countermeasure list with priority ranking	Ref to "overall priority" column in form 6.

[a]Forms in this column refer to the forms in the API SRA standard document [10]

countermeasures are ranked according to their potential risk reduction ΔR_{cm} as well as some other practical information (e.g., costs).

$$\Delta R_{cm} = \sum\nolimits_{a \in CAL} \sum\nolimits_{t \in TL} \sum\nolimits_{s \in Sce(a,t)} \left(R^2_{(a,t,s,cm)} - R^1_{(a,t,s)} \right). \tag{1.1}$$

The API SRA methodology is a systematic process that (i) evaluates the success-ful likelihood of a threat against a facility, (ii) considers the potential severity of consequences to the facility itself, to the surrounding community, and to the energy supply chain, (iii) provides clear definitions of terminologies used in the methodol-ogy, (iv) clearly points out what inputs the methodology needs for the security risk assessment procedure and what outputs the methodology will provide, and (v) gives guidance on how to organize a SRA team.

Focusing on minimizing the defender's maximal loss and taking into account uncertainties during the security risk assessment procedure, the API SRA method-ology would output robust results. Though it is not a fully quantitative risk assess-ment based methodology, it is performed qualitatively using the best judgment of the SRA Team. Comparing to the SRFT, the API SRA methodology is more concrete to execute, and considers not only the facility itself, but its surroundings as well.

The API SRA methodology suffers two drawbacks. First, the methodology fails to model the dynamic (intelligent) interactions between defender and attackers. As shown in Fig. 1.6, the SRA team estimates the attractiveness of each asset to each threat at the very beginning of the procedure. However, after presuming that the recommended countermeasures are implemented, the SRA team does not re-estimate the attractiveness. Therefore, in reality, the attackers would change their targets according to the defender's plan. Second, risk scoring methods and risk matrices are employed in the API SRA methodology. For example, Cox [27] and Baybutt [28] have criticized the use of risk scores and risk matrices and proposed improvements.

1.3.4   Drawbacks of Current Methodologies

The current methodologies used for security risk assessment within the chemical and process industry mainly suffer from two drawbacks: (i) they are all qualitatively based and (ii) they fail to model intelligent interactions between the defender and the attacker.

Qualitative models can only inform industrial managers about which part of the plant needs to be better protected and it does not mention how many improvements are needed. Ideally, the defender needs quantitative guidance to make decisions, such as how to allocate the limited security resources. Qualitative models can also be theoretically not sound. For instance, Cox [29] lists several theoretical limitations of the security risk assessment methodologies which are based on the "risk = threat × vulnerability × consequence" formula. Zhang et al. [30] suggests several further impediments that the API SRA methodology needs to pay more attention to.

Despite the drawback of models being qualitative, being not able to model dynamic interactions between the defender and the attacker is the most important and essential problem of the above mentioned conventional security risk assessment methods. As also mentioned by Baybutt [1], these conventional security methods are mostly derived from safety risk management methods and can be compared with PHAs. Therefore, security risks are calculated by using a "risk = probability × consequence" approach. However, adversaries related to security risks are to be considered as 'intelligent opponents', thus not acting randomly or probabilistically. Instead, intelligent attackers behave according to their goal and also based on the difficulties of reaching their goal. See in Fig. 1.2 that how the security methodology without considering intelligent attackers can mislead allocation of security resources.

1.4 Protection of Chemical Industrial Parks (CIPs) or So-Called Chemical Clusters

1.4.1 Security Within Chemical Clusters

As Curzio and Fortis [31] state, firms decide to settle in a cluster on the basis of the expected profitability of being located there. This profitability depends on geographical and agglomeration benefits, obtained as the difference between gross location-related benefits and costs. As the number of corporations located in an industrial cluster increases, gross benefits increase due to productive specialization, scientific, technical and economic spillovers, reduction in both transport and transaction costs, increases in the quality of the local pool of skilled labour force, etc. This observation also explains why chemical plants form chemical clusters. However, in the case of chemical enterprises, clustering not only implies profit opportunities and economic benefits of scale. A chemical cluster has a very high responsibility towards maintaining safety and security standards in the urban surroundings as well. Each additional chemical plant entering a chemical cluster might decrease the average safety and security standing of the area.

Companies in chemical clusters are thus not merely linked via technological spillovers, logistics advantages, and so on. They are related through the responsibility of looking after safety and security requirements in the entire cluster as well. Thus, safety and security does not stop at the companies' fences, on the contrary, a terrorist attack can be easily deliberately inducing an accident within a company with cross-border consequences, and may cause even more severe accidents in nearby companies. Such scenarios, although very much possible, are currently not taken into account, nor by the legislator, nor by the industry. Nevertheless, as terrorism can be defined as the unlawful use of – or threatened use of – force or violence against individuals or property to coerce or intimidate governments or societies, often to achieve political, religious or ideological objectives [32], it is obvious that chemical clusters, wherever situated worldwide, may form an important target to terrorist activities.

As mentioned before, approaches to tackle security risks, are well-known and widely used amongst security experts. However, current security assessment approaches are static and they fail to capture intelligent interactions and systemic risks. Due to the latter weakness, they are rather myopic in their possible use. Countermeasures against terrorist threats based on current vulnerability assessments are thus limited to static immediate consequences and are neither designed to prevent an accident attaining systemic proportions or intelligent interactions, nor to effectively limit the consequences of a large-scale (for instance multi-target) terrorist attack.

Security risk assessment methodologies that aim to assess risks at a higher level, e.g. for networked systems, therefore require further refinement. A detailed risk and vulnerability assessment at this higher level is no longer applicable and a certain level of abstraction is necessary.

Some representative examples of large chemical clusters are the chemical industrial area in the Port of Antwerp in Belgium, the Rotterdam Port chemical cluster in the Netherlands, the Houston chemical industrial park, the Shanghai chemical industrial area, etc. Obviously, chemical plants in one park can share some infrastructure, such as perimeters, transportation stations etc. Clustering plants together and locate them with a certain distance from populated areas is an efficient and economic choice for governments. However, as indicated above, being clustered also increases risks for chemical plants. Due to the existence of domino effects [3], an accident in one plant may also spread to its adjacent plants. Therefore, plants in a cluster not only have to deal with risks within the plant, but also they have to take the shared risks (external risks) into account. This is expounded more clearly in the next section. Figure 1.7 shows an example of the shared risks in a chemic cluster, in which the arrows denote the domino effects from the source hazardous installation to the target hazardous installation. In Fig. 1.7, arrows "De2" and "De5" represent risks of companies resulting from their neighbour companies. Domino effects occurred in a number of accidents in the process industries. A list of these accidents can be found in Reniers [33] and Reniers and Cozzani [3].

1.4.2   Chemical Cluster Security: State-of-the-Art Research

As already indicated, although a lot of knowledge and know-how has been built up in the chemical industry as regards process safety, safety practices and to create adequate safety cultures within chemical corporations, process security is a rather new domain, which largely gained interest from regulators, practitioners and academics after the 9/11 attacks in New York. At present, operational security in the

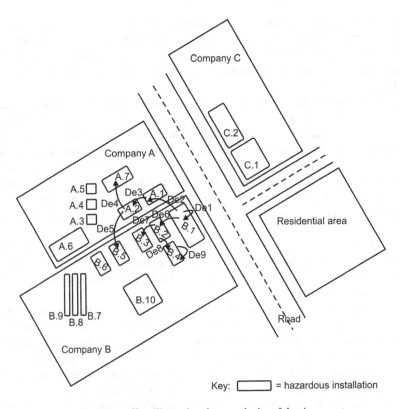

Fig. 1.7 Hypothetical domino effect illustrating the complexity of domino events

process industries seems largely to be legislation-driven. However, similar to process safety, process security within a chemical plant should be a choice of management and should be seen as related with a management conviction of the need to avoid losses, in order to establish a profitable company. In this regard, top management should be more open-minded as regards terrorism prevention. Chief Executive Officers usually overestimate the perceived costs of security measures for Type II events, and underestimate their potential benefits. Consequently, short-term decisions are taken. A way to make more rational investments in terrorism prevention is to be open to any new information, and to examine what kind of hypothetical benefits (gained by avoiding a terrorist attack) could be derived from large-scale security investments. It would thus be recommendable that security management provides top management with detailed security cost/benefit figures. Such figures should more objectively demonstrate the requirements of the chemical company as regards counter-terrorism security. The formulation of attack scenarios and the costs and benefits associated with thwarting such attacks should be soundly conducted. This dedication and the ability to conceive the impossible or the unthinkable should demonstrate that the chemical company and all its employees benefit from a plant security culture and from installing, maintaining and continuously improving counter-terrorism measures.

Furthermore, process security should be looked upon from the level of the industrial cluster, and not only from the chemical plant level. Single chemical plants are still focusing too much on individual efforts and advancements for becoming more secure. If situated in a cluster, chemical plants require foremost a holistic collaborative approach and the optimal use of existing knowledge and know-how on an industrial park level. Industrial symbiosis initiatives within chemical industrial areas throughout Europe are currently concentrated on operational collaboration as regards linked production, linked delivery of services and/or related with the open innovation concept. Collaboration on a strategic level to pro-actively enhance process security within chemical industrial parks has however not yet been sufficiently explored and exploited.

Nonetheless, chemical installations are mutually linked in terms of the level of danger they pose to each other, irrespective of company fences. One type of accident particularly interesting in this regard, is a cross-company escalating accident or a so-called (external) domino effect. Such accidents may theoretically even affect an entire chemical industrial park, and thus represent "systemic risks" within the chemical industry. These risks may be accidental, i.e. safety-related, but they may also be intentionally induced, that is, security-related. To deal with such risks, intensive pro-active collaboration and joint efforts on different levels (strategic, tactic and operational) are needed in chemical clusters, and intelligent attacks need to be considered in this "collaboration against terrorism". See for instance, Reniers and Soudan's work on a game theoretical approach for reciprocal security-related prevention investment decisions [34]. Thus, the traditional single-chemical-plant approach of dealing with security risks in chemical parks needs to be complemented by a 'systemic risk approach'. We have only recently discovered that chemical clusters, mathematically seen as networks of installations linked via dangers of potential escalation, might follow a power-law distribution [35]. Hence, intelligent attenuation-based security could for example be introduced in such parks.

1.4.3 Future Promising Research Directions on Cluster Security

It is a logical evolution that safety and security within chemical plants follow the same bottom-up approach as present in almost all other aspects of nature, industry, and society. Safety and security started within individual chemical companies during the last decades of the previous century, but now the time has come that safety and security reach the level of the chemical industrial park. Therefore, it is imperial that the governance of safety and security arrangements in chemical clusters are thoroughly studied (e.g., how to share costs and decision-making, how to pool resources), and the accompanying policy and regulatory dimensions are explored in depth (e.g., the role for government, federal or local, in providing incentives and resources). Furthermore, strategic safety and security collaboration leads to competitive advantages due to spillovers, trust-increasing effects, and decreased accident likelihood within the cluster.

From the above, it is clear that different from safety research that focuses on natural and randomly (non-strategic) hazards, security research has to face intelligent and strategic adversaries. Traditional methods or concepts used in safety science such as probabilistic risk assessments, historical data analysis and what have you, no longer can be readily and easily used in security research. When dealing with security problems, the adversaries' strategies should be taken into consideration instead of the incidents' probabilities.

Game theory, which originated in economic sciences, is a good choice to handle problems that contain intelligent players. Game theory has very rigor mathematical foundations, and if adequately used with respect to chemical security, we can obtain more accurate and more defensible quantitative results, besides the qualitative assessments and results used nowadays in chemical plant security management. In recent years, a lot of attention in academia has been laid on the combination of game theory and critical infrastructure protection. Tambe and his group [36] used game theory to improve the security situation in airport patrolling, air marshals' allocation, and coast line protection. They developed several decision support systems based on their research, and these systems now work in reality. Bier and her group [37] studied the combination of game theory and security assessment methods from a theoretical viewpoint. They answered the questions why game theory has an important role in security research, and illustrated the advantages and disadvantages of using game theory in operational security.

Although there are already some researches on using game theory to improve operational security, in fact in the chemical process security, very scarce research has been done as yet. Security problems in the process industries are different to those in aviation or the electric power grid for example, although they are all critical infrastructures. We cannot readily apply game theoretical models now being used in aviation, within the process industries directly. Different security models are implemented in different types of industries. For instance, in Tambe's model, air marshals are allocated to defend an air plane, and therefore the players' strategies are limited to "protect" (that is, to allocate an air marshal on the plane) or not (that is, no air marshal on the plane), and "attack" or not. However, in case of security in the process industries, the model is more complex, the strategies may be at a different alert level (discrete model) or at a different investment level (continuous model).

1.5 Conclusion

The protection of single chemical plants as well as the protection of small and large chemical clusters have been an important task for risk analysts. The chemical industry, on the one hand fulfils an extremely important role for our modern lives, but on the other hand it poses huge threat to modern society. If installations storing toxic, flammable, or explosive materials would be damaged by intentional attacks, the consequences would be awful. Moreover, the two attacks on chemical plants in June and July 2015 in France proved the possibility of an attack to the chemical industry in the Western world.

There are plenty of academic studies concerning the protection of chemical installations. Also, regulations, standards, and guidelines on promoting chemical security have been published, especially in the U.S. However, due to the lack of historic data and the failure to model intelligent interactions between the malicious attackers and the defenders, the current researches and regulations etc. have their drawbacks. Moreover, there is a lack of effort with respect to the protection of chemical clusters, which, if being strategically attacked, may result in truly catastrophic consequences for society.

Game theory, being able to support strategic decision making, has been successfully applied in several domains for improving security. Hall Jr [38] (2009) mentions that "If the conditions creating the problems you had to deal with were natural or random, the answer was decision analysis (which looked a lot like what we now call risk analysis). If the conditions creating the problems you had to deal with were the result of deliberate choice, the answer was game theory." Therefore, we conclude that game theory has the potential to be a proper methodology for improving security in the chemical and process industries.

References

1. Argenti F, Bajpai S, Baybutt P, Cozzani V, Gupta J, Haskins C, et al. Security risk assessment: in the chemical and process industry. Berlin: Walter de Gruyter GmbH & Co KG; 2017.
2. Database GT. 2016. Available from: https://www.start.umd.edu/gtd/
3. Reniers G, Cozzani V. Domino effects in the process industries: modelling, prevention and managing. Amsterdam: Elsevier B.V; 2013. 1–372 p
4. (DHS) DoHS. National strategy for homeland security. 2002.
5. Anastas PT, Hammond DG. Inherent safety at chemical sites: reducing vulnerability to accidents and terrorism through green chemistry. Amsterdam: Elsevier; 2015.
6. Luft G. Pipeline sabotage is terrorist's weapon of choice. Pipeline Gas J. 2005;232(2):42–4.
7. Reniers G, Pavlova Y. Using game theory to improve safety within chemical industrial parks. London: Springer; 2013.
8. Reniers G, Khakzad N. Revolutionizing safety and security in the chemical and process industry: applying the CHESS concept. J Integr Secur Sci. 2017;1(1):2–15.
9. Powell R. Defending against terrorist attacks with limited resources. Am Polit Sci Rev. 2007;101(03):527–41.
10. API. Security risk assessment methodology for the petroleum and petrochemical industries. In: 780 ARP, editor; 2013.
11. Guikema SD, Aven T. Assessing risk from intelligent attacks: a perspective on approaches. Reliab Eng Syst Saf. 2010;95(5):478–83.
12. Meyer T, Reniers G. Engineering risk management. Berlin: Walter de Gruyter GmbH & Co KG; 2016.

13. Baybutt P. Analytical methods in process safety management and system safety engineering–process hazard analysis. Handb Loss Prev Eng. 2013;1&2:501–53.
14. Baybutt P. Assessing risks from threats to process plants: threat and vulnerability analysis. Process Saf Prog. 2002;21(4):269–75.
15. Dunbobbin BR, Medovich TJ, Murphy MC, Ramsey AL. Security vulnerability assessment in the chemical industry. Process Saf Prog. 2004;23(3):214–20.
16. Bajpai S, Gupta J. Site security for chemical process industries. J Loss Prev Process Ind. 2005;18(4):301–9.
17. Bajpai S, Gupta J. Securing oil and gas infrastructure. J Pet Sci Eng. 2007;55(1–2):174–86.
18. Garcia ML. Vulnerability assessment of physical protection systems. Amsterdam: Butterworth-Heinemann; 2005.
19. Nolan DP. Safety and security review for the process industries: application of HAZOP, PHA, What-IF and SVA Reviews. Waltham: Elsevier; 2014.
20. Baybutt P. Security vulnerability analysis: protecting process plants from physical and cyber threats. In: Reniers G, Khakzad N, Gelder PV, editors. Security risk assessment: in the chemical and process industry, vol. 1. Berlin: Walter de Gruyter GmbH & Co KG; 2017.
21. Kaplan S. The words of risk analysis. Risk Anal. 1997;17(4):407–17.
22. Baybutt P. Get creative with process safety management. Chem Eng Prog. 2017;113(4):58–60.
23. Baybutt P. A framework for critical thinking in process safety management. Process Saf Prog. 2016;35(4):337–40.
24. Moore DT. Critical thinking and intelligence analysis. Washington, DC: National Defense Intelligence Coll; 2007.
25. Baybutt P. Cognitive biases in process hazard analysis. J Loss Prev Process Ind. 2016;43:372–7.
26. Montibeller G, Winterfeldt D. Cognitive and motivational biases in decision and risk analysis. Risk Anal. 2015;35(7):1230–51.
27. Cox LAT, Babayev D, Huber W. Some limitations of qualitative risk rating systems. Risk Anal. 2005;25(3):651–62.
28. Baybutt P. Designing risk matrices to avoid risk ranking reversal errors. Process Saf Prog. 2016;35(1):41–6.
29. Cox LAT Jr. Some limitations of "Risk = Threat × Vulnerability × Consequence" for risk analysis of terrorist attacks. Risk Anal. 2008;28(6):1749–61.
30. Zhang L, Reniers G, Chen B, Qiu X. Integrating the API SRA methodology and game theory for improving chemical plant protection. J Loss Prev Process Ind. 2018;51(Supplement C):8–16.
31. Curzio AQ, Fortis M. Complexity and industrial clusters: dynamics and models in theory and practice. Heidelberg: Springer; 2012.
32. Gulak M, Kun U, Koday Z, Koday S. Preventing terrorist attacks to critical. Underst Respond Terrorism Phenom A Multidimens Perspect. 2007;21:298.
33. Reniers GLL. Multi-plant safety and security management in the chemical and process industries. Weinheim: Wiley-VCH; 2010.
34. Reniers G, Soudan K. A game-theoretical approach for reciprocal security-related prevention investment decisions. Reliab Eng Syst Saf. 2010;95(1):1–9.
35. Reniers GLL, Sörensen K, Khan F, Amyotte P. Resilience of chemical industrial areas through attenuation-based security. Reliab Eng Syst Saf. 2014;131:94–101.
36. Tambe M. Security and game theory: algorithms, deployed systems, lessons learned. Cambridge: Cambridge University Press; 2011.
37. Bier VM, Azaiez MN. Game theoretic risk analysis of security threats. Dordrecht: Springer; 2008.
38. Hall JR Jr. The elephant in the room is called game theory. Risk Anal. 2009;29(8):1061.

Chapter 2
Intelligent Interaction Modelling: Game Theory

2.1 Preliminaries of Game Theory, Setting the Scene

2.1.1 Introduction

Game theory is a mathematical tool for supporting decision making in a multiple players situation where one player's utility will be determined not only by his own decision, but also by other players' decisions. An illustrative example of this situation is the Rock/Scissors/Paper game ("RSP" game). In an RSP game, whether a player wins or loses depends on both what he plays and what his opponent plays. This is a well-known game between mostly children with very simple rules. Two 'players' hold their right hands out simultaneously at an agree signal to represent a rock (closed fist), a piece of paper (open palm), or a pair of scissors (first and second fingers held apart). If the two symbols are the same, it's a draw. Otherwise rock blunts scissors, paper wraps rock, and scissors cut paper, so the respective winners for these three outcomes are rock, paper and scissors. The RSP game is what is called a 'two-player zero-sum non-cooperative' game. There are obviously many other types of game and the field of game theory is very powerful to provide (mathematical) insights into strategic decision-making.

Game theory was formulated as a research domain after von Neumann and Morgenstern's work [1]. Before their work, there was scattered research on interactive decision making, in which the idea of game theory was employed. Among others, Cournot's duopoly model, for example, studied how to predict the production of two monopolistic companies. The Stackelberg leadership model, on the other hand, investigated how to predict production of different companies when there is a leader/dominant company. von Neumann and Morgenstern [1] systematically studied strategic behaviours in the economic area, and proposed the famous MaxiMin theory based on a zero-sum game. Nash [2] studied general sum games, and proved that in a game with finite players and finite strategies, a Nash equilibrium always

© Springer International Publishing AG, part of Springer Nature 2018
L. Zhang, G. Reniers, *Game Theory for Managing Security in Chemical Industrial Areas*, Advanced Sciences and Technologies for Security Applications,
https://doi.org/10.1007/978-3-319-92618-6_2

exists. Harsanyi [3] investigated games with incomplete information, and proposed the Harsanyi transformation to transfer an incomplete information game to a complete but imperfect information game. In the twentieth century, game theoretic research is mainly stimulated by economists and mathematicians, and several game theorists were awarded the Nobel prize, such as John Nash, Robert Aumann, and Lloyd Shapley etc. Furthermore, actually, all five game theorists who have won Nobel Prizes in economics, have been employed as advisors to the U.S. Pentagon at some stage in their careers.

Since the end of the twentieth century, with the advances in computer science and the power of computer technology, game theory has been introduced to the computer science community. In the application perspective, game theory can be used for the allocation of network resources, for the modelling of intelligent agents in the artificial intelligence domain, for adversarial machine learning etc. Some computer scientists focus on theoretically developing efficient algorithms to calculate equilibria for large-scale game theoretic models. It is worth noting that Nash proved the existence of Nash equilibrium (NE) (see also Sect. 2.2.2) in finite games, as mentioned above, however, his proof is not a constructive proof. Therefore, algorithms for computing the NE must be developed. Lemke and Howson [4] proposed an algorithm for searching one NE in a bi-matrix game. Chen and Deng [5] further proved that the task of computing a NE in a two-player game cannot be finished in polynomial time. Interested readers for computational issues in game theory are referred to Nisan et al. [6]

As made clear before, a central feature of multi-person interaction is the potential for the presence of strategic interdependence. The actions which are best for one decision-maker may depend on actions which other individuals have already taken, or are expected to take (or not take). The tool that we use for analysing interactions with strategic interdependence is non-cooperative game theory. The term 'game' actually highlights the theory's central feature: the decision-makers under study are concerned with strategy and winning (in the classic micro-economic sense of utility- or profit maximization). The decision-maker will have some control over the situation, but not all control since other decision-makers' actions also influence the outcome.

Basically, a game theoretic model consists of players (that is, decision-makers), strategies, and payoffs. Two assumptions, namely the 'common knowledge' assumption and the 'rationality' assumption, are often discussed in game theoretic models. Furthermore, different game solutions need to be employed for simultaneous games and for sequential games.

2.1.2 Players

Players need to be seen as strategic actors involved in the game. Actors can be people, but also institutions, organisations, etc., and even countries. A game theoretic model must contain at least two players. If there are only two players, the game is called a two-player game, otherwise the game is called a multiple-player game.

If cooperation can be achieved between players, the game is called a cooperative game, otherwise it is called a non-cooperative game. For instance, in a chemical cluster security case, different plants may cooperate with each other to jointly strategically invest in security measures, and thus a cooperative game can be applied to model this situation. The defender (cluster organisation) and the potential attacker (e.g., terrorists), however, will never cooperate of course, and thus a non-cooperative game should be employed in this case.

In this book we mainly focus on two-player non-cooperative games.

2.1.3 Strategy (Set)

Each player in a game in principle has a set of strategies. A strategy set defines the player's behaviour rules when playing the game. A behaviour rule means that the player chooses a certain action at a certain step in the game (in game theory terminology this is called an information set). Therefore, a strategy can be seen as a series of actions of a player.

For instance, in a simple defender-attacker game, assume that the defender (the attacker) only has two actions, namely, "not defend (no attack)" and "defend (attack)". Figure 2.1 (left-hand side) shows the game tree when the attacker does not know the defender's action yet when he has to make a decision (i.e., the game is played simultaneously, see also Sect. 2.1.7). Figure 2.1 (right-hand side) shows the game tree when the defender plays first and the attacker plays afterwards knowing what action the defender has played (i.e., the game is played sequentially, see also Sect. 2.1.7). In both cases, the defender's strategy set equals her action set, that is:

$$S_d = \{\text{ND : not defend}, \text{D : defend}\}.$$

In the simultaneous game, the attacker's strategy set equals his action set, being the following couple of strategies:

Fig 2.1 Game tree of a illustrative defend-attack game

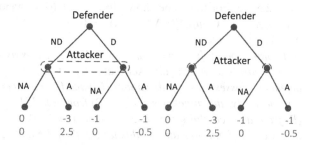

$$S_a = \{NA : \text{no attack}, A : \text{attack}\},$$

while in the sequential game, the attacker's strategy set is different from his action set, that is:

$$S_a = \begin{cases} NA - NA : \text{if ND, then NA, if D, then NA; NA} \\ -A : \text{if ND, then NA, if D, then A; A} - NA : \text{if ND, then A, if D, then NA;} \\ A - A : \text{if ND, then A, if D, then A} \end{cases}$$

In the simultaneous game as shown in Fig. 2.1 (left-hand side), the attacker only has one information set, which is shown as a dotted line oval. However, in the sequential game, the attacker has two information sets, which is shown as a circle in Fig. 2.1 (right-hand side). The attacker's strategy set should define the rules how the attacker could move at each information set, hence resulting in four strategies in the sequential case.

The strategy set we have defined is also called the "pure strategy set". A so-called 'mixed strategy' allows the player to probabilistically play each pure strategy [7]. In the latter case, the sum of the probabilities that each pure strategy is played, equals to 1. For instance, in the illustrative game shown in Fig. 2.1, the mixed strategy space for the defender can be defined as:

$$X = \{x \in R^2 | x_{ND}, x_D \geq 0, x_{ND} + x_D = 1\},$$

in which x_{ND} represents the probability that the pure strategy ND will be played and x_D denotes that probability that the pure strategy D will be played.

2.1.4 Payoff

The payoff of a game models (and measures) the player's interests/goals/aims/ preferences in the game. A payoff needs to be defined for each player and for every combination of strategies, that is, $u_i\left(\prod_{p=1}^{N} s_p\right) \rightarrow R$, in which, N represents the number of players in the game, s_p denotes the pth player's pure strategy, u_i denotes the ith player's payoff. For instance, in the game shown in Fig. 2.1 (left-hand side), we have $u_D(ND, NA) = 0$, $u_D(ND, A) = -3$, $u_A(ND, NA) = 0$, $u_A(ND, A) = 2.5$ etc. For the game shown in Fig. 2.1 (right-hand side), we have $u_D(ND, NA - NA) = 0$, $u_D(ND, NA - A) = 0$, $u_D(ND, A - NA) = -3$, $u_D(ND, A - A) = -3$ etc.

Explanation of the payoff numbers in Fig. 2.1: It is assumed that the defender's defence cost (C_d) and the attacker's attack cost (C_a) are 1 and 0.5 respectively; the defender's losses from a failed attack (L_0) and from a successful attack (L) are 0 and 3 respectively; the attacker's penalty (P) from a failed attack is 0 while his reward (R) from a successful attack is 3. Further assume that if the defender defends the target and the attacker attacks this target, then the attack will fail, otherwise if the *defender does not defend the target and the attacker attacks this target, then the attack will succeed. Therefore, if the defender plays 'ND' and the attacker plays 'NA', then both players have a payoff of 0; if the defender plays 'ND' and the attacker plays 'A', then the attack will succeed and the defender loses her value from the target (i.e., −3) and the attacker gets his reward but loses his attack cost (i.e., 3–0.5 = 2.5); if the defender plays 'D', and the attacker plays 'NA', then the defender loses her defence cost (i.e., −1) and the attacker has a payoff of 0; if the defender plays 'D' and the attacker plays 'A', then the attack will fail and the defender only loses her defence cost (i.e., −1) and the attacker loses his attack cost and suffers a penalty (i.e., −0.5–0 = −0.5).*

A zero-sum game is a game that under every combination of the players' pure strategies, the sum of all the players' payoffs equals zero, i.e., $\sum_{i=1}^{N} u_i \left(\prod_{p=1}^{N} s_p \right) = 0$. A game is called a non-zero-sum game, if there exists such a combination of players' pure strategies $\prod_{p=1}^{N} s_p$ that $\sum_{i=1}^{N} u_i \left(\prod_{p=1}^{N} s_p \right) \neq 0$. Zero-sum games have some special properties, such as being easier to solve. It is worth noticing that the idea of duality theory in linear programming was originally developed from the study of two-player zero-sum games.

In a two-player game where each player has a finite set of pure strategies, if player 1 and 2 play mixed strategies x and y respectively, then the expected payoff for the players can be calculated as shown in Formulas (2.1) and (2.2).

$$EU_1 = \sum_{i=1}^{m} \sum_{j=1}^{n} x_i \cdot U_1(i,j) \cdot y_j \qquad (2.1)$$

$$EU_2 = \sum_{i=1}^{m} \sum_{j=1}^{n} x_i \cdot U_2(i,j) \cdot y_j \qquad (2.2)$$

In which EU_1 and EU_2 are the expected payoffs for player 1 and for player 2 respectively; m and n denote the number of pure strategies of player 1 and player 2 respectively; U_1 and U_2 are the payoff matrices for player 1 and player 2 respectively. Both U_1 and U_2 have m rows and n columns, and entries (i,j) of these two matrices represent the player's payoff when player 1 plays his i^{th} pure strategy and player 2 plays her j^{th} pure strategy.

2.1.5   The Assumption of 'Common Knowledge'

The 'common knowledge' assumption in a game theoretic model assumes that each player in the game has global information of the game. A game where this is the case is also called a 'complete information' game. In such a game, players know their own strategies and payoff functions and also those of the other players. In addition, each player knows that the other players have complete information. In reality, this is a very strong, perhaps in some cases much too strict, assumption. Nonetheless, research suggests that good managers are very well informed, multi-skilled and flexible in their approach to problem-solving. Organisations themselves are increasingly complex places, which by no means live in isolation anymore from the expectations of other micro-economic players (including opponents such as competitors and terrorists). Knowledge and know-how, mathematical or otherwise, such as explained in this book from a game theoretical perspective, is often what separates a failing manager from a successful one.

For instance, in an RSP game, every player knows the rules of the game. In the illustrative game shown in Fig. 2.1, a common knowledge assumption means that the figure is known to both the defender and the attacker, and they know whether the game is played simultaneously or sequentially.

The common knowledge assumption can also be interpreted in iterative way:

1. the players know their own information (i.e., strategies and payoffs) and know how the game will be played;
2. the players also know all other players' information;
3. the players know that other players know their information;
4. and so forth.

The common knowledge assumption is satisfied in many famous games, such as in the rock/scissors/paper game, the chess game, and the Go game. Conversely, in most poker games (e.g., the Texas hold'em), the common knowledge does not hold.

In the security domain, the defender and the attacker are often modelled as the two players in the game. Defenders even have difficulties to exactly know their own information, for instance, to know how severe an attack could be. Therefore, the common knowledge assumption is rather a strong requirement for security game modellers.

Rios and Rios Insua [8] proposed an adversarial risk analysis (ARA) approach for *relaxing* the common knowledge assumption in the security game models. In the ARA framework, all the data are estimated by the defender. And she:

(i)   knows her own data u_d^0;
(ii) estimates the attacker's data $u_a^0 {\sim} F_1$, uncertainties distribution F exist in this step and afterwards since the defender would not be able to know the exact numbers of these data;
(iii) estimates the attacker's estimation of the defender's data $u_d^1 {\sim} F_2$;
(iv) estimates the attacker's estimation of the defender's estimation of the attacker's data $u_a^1 {\sim} F_3$;
(v)   ...
(vi) At a certain step, presumes that the attacker behaves randomly.

This iterative estimation of data results in the fact that in a security game (or, in the ARA framework), the defender and the attacker are both behaving according to their own data as well as according to (what they think will be) their opponent's decision. In step (i), the defender knows her own data. However, her optimal decision should also be influenced by the attacker's decision. Therefore, she has to move to step (ii). In step (ii), the defender has an estimation of the attacker's data. However, the attacker's decision should also be influenced by the defender's decision, thus the attacker's estimation of the defender's data is required, as stated in step (iii). In theory, this iterative step will go deeper infinitely. In modelling practise, at a certain step, the defender presumes that the attacker plays randomly. Knowing the attacker's behaviour, the defender can then stop this data iteration, and move backwards by calculating both players' optimal strategies until step (i) again.

The ARA framework theoretically relaxes the common knowledge assumption. However, it needs massive inputs (i.e., the u_d^0, F_1, F_2...) and these inputs are also difficult to obtain. Nonetheless, in a defender-attacker game, if the attacker would know the defender's strategy when he moves, the iteration would be stopped at step (ii). At step (ii), the defender knows the attacker's data $u_a^0 {\sim} F_1$ and she also knows that the attacker can observe her strategy, instead of guessing her strategy. Therefore, the defender can work out the attacker's behaviour, and then move back to step (i), and combine this information with her own data, to play optimally.

In games of 'incomplete information', the common knowledge assumption should not hold, and players know the rules of the game and their own preferences, but not the payoff functions of the other players.

To have a complete notion of the different possible games (and game terminology) and an overview of the games in this section, we need to mention also games of 'perfect information' and games of 'imperfect information'. The former type of games are those in which players select strategies sequentially and are aware what other players have already chosen, such as the game of chess. The latter type of games represent those in which players have to act in ignorance of one another's moves, merely anticipating what the other player will do.

2.1.6 The Assumption of 'Rationality'

The assumption of rationality indicates that people always behave in their own best interests. A player with full rationality thus means that he/she is playing to maximize his/her own payoff. As argued by Kelly [9], the assumption of rationality can be justified on a number of levels. At its most basic level, it can be argued that players behave rationally by instinct. However, experience suggests that this is not always the case, since decision makers frequency adopt simplistic algorithms which very often lead to sub-optimal solutions. Secondly, it can be argued that there is a kind of natural selection at work which inclines a group of decisions towards the rational and optimal. Successive generations of decisions are increasingly rational. Finally, it has been suggested that the assumption of rationality that underpins game theory is not an attempt to describe how players actually make decisions, but merely that they behave *as if* they were not irrational [10]. All theories and models are, by definition, simplifications and should not be dismissed simply because they fail to represent all realistic possibilities. A model should only be discarded if its predictions are false or useless, and game theoretic models are neither [9].

Allowing for the possibility that there are several equally attractive best actions, the theory of rational choice boils down to the following definition: *the action chosen by a player is at least as good, according to his/her preferences, as every other available action.* Standard micro-economic theories of the consumer and the company also use this assumption as one of the basic rules for applying the theories.

The assumption of rationality is a very important component of models and theories that lead to our understanding of social phenomena. Nonetheless, under some conditions its implications are at variance with observations of human decision-making, as Osborne [11] explains. To take a small example, adding an undesirable action to a set of actions sometimes significantly changes the action chosen.

Near-rationality, or so-called 'bounded rationality', allows players to be rational, but only within certain limits. Players are allowed to play sub-optimal strategies as long as the payoff per iteration is within a certain (small) positive number of their optimal strategy. For instance, Pita et al. [12] studied a so-called 'epsilon-optimal' player in their security games. An 'epsilon-optimal' attacker is an attacker who would deviate from his optimal strategy to strategies that have close payoff to the payoff that he can obtain from the optimal strategy.

2.1.7   Simultaneous and Sequential Game

As already mentioned, essentially two types of games are possible: (i) games where the moves of the players cannot be seen by the other players, hence these are hidden-move games which are also called 'simultaneous-move games', and (ii) games where the players make moves in some sort of order, hence these are transparent games which are also called 'sequential-move games' or dynamic games.

The RSP game we mentioned in the beginning of this chapter, for instance, is a classic simultaneous game. Most table games, such as the chess game, the go game, and the Texas hold'em etc., are typical sequential games, since in these games, players are choosing their strategies with knowing their opponents' chosen strategies.

It is worth noting that the temporal order of choosing strategies does not determine whether a game is a simultaneous game or a sequential. Instead, only in case that when some players play first (being game leaders) and other players can observe these game leaders' played strategies (being game followers), the game is a sequential game. For instance, in the illustrative defend-attack game we discussed in Sect. 2.1.3, the defender always moves first by deciding whether to defend or not, while the attacker follows by whether to attack or not. However, if the attacker cannot observe the defender's played strategy (i.e., either 'not defend' or 'defend'), then the game is a simultaneous game and the game tree is shown on the left-hand side of Fig. 2.1. If the attacker knows the defender's played strategy when he makes his

Table 2.1 Strategic form of the simultaneous move game for the illustrative defend-attack game

		Defender	
		ND	D
Attacker	NA	0,0	0,–1
	A	2.5,–3	–0.5,–1

Table 2.2 Strategic form of the sequential move game for the illustrative defend-attack game

		Defender	
		ND	D
Attacker	NA-NA	0,0	0,–1
	NA-A	0,0	–0.5,–1
	A-NA	2.5,–3	0,–1
	A-A	2.5,–3	–0.5,–1

decision, then the game is a sequential game and the game tree is shown on the right-hand side of Fig. 2.1. Tables 2.1 and 2.2 further show the strategic forms for the simultaneous move game and for the sequential move game, respectively, of the illustrative defend-attack game.

2.2 Game Theoretic Models with a Discrete Set of Strategies

2.2.1 Discrete and Continuous Set of Strategies

A player's strategy set can either be discrete or continuous, depending on the modelling approach. In the defend-attack game illustrated in Sect. 2.1.3, the defender's strategies were modelled as either "Defend" or "Not Defend" and the attacker's strategies were modelled as either "Attack" or "Not Attack", both thus being discrete sets of strategies. These discrete sets of strategies can be understood as defend and attack scenario based. For instance, the defender has only one security measure at her disposal, and she can decide whether to implement this measure to protect the target (i.e., "Defend") or not (i.e., "Not Defend"). Similarly, the attacker decides whether to use a specific attack scenario (e.g., a Vehicle-Born Improvised Explosive Device) or not.

Strategies of the game illustrated in Sect. 2.1.3 can also be continuous. The players' strategies can be their defence and attacker efforts, for the defender and for the attacker respectively. These continuous strategies can be interpreted as resources-based. For example, the defender may have a maximal amount of security budget B, and she can allocate arbitrary money $b \in [0, B]$ to the target. The attacker's continuous strategies can also be interpreted in a similar way.

In this book, we focus on game models with discrete strategy sets, since game theoretic models with continuous strategy sets suffer several drawbacks.

First of all, it is theoretically difficult to solve a game model with continuous strategy sets. Analytical approaches should be employed to solve these games. If the game becomes more complicated, there may be many variables and the difficulties of using analytical methods might increase dramatically. In the illustrative game in Sect. 2.1.3, if there would be multiple targets – which is always the case in reality – and if the vulnerability of the target is a non-linear function of the defence effort and the attacker effort (e.g., by using a contest success function [13] – which is often the case in reality – then a non-linear optimization problem with multiple variables will have to be solved in order to calculate the solution for the game. Furthermore, solutions are not guaranteed in these games, since the most famous solution of game theoretic models, which is, the Nash Equilibrium, is only guaranteed in a game with finite players and finite pure strategies [2]. Conversely, if the game would be modelled with discrete strategy sets, a Nash Equilibrium always exists. The computation of a Nash Equilibrium in a finite security game is also difficult, as we stated in the beginning of this chapter. However, after Lemke and Howson's work [4], a lot of researchers have improved the algorithms that are capable to solve a finite game. Moreover, the computational capability of computers has improved dramatically over the past decades, and it keeps improving. Nowadays, we are able to solve game theoretic models with thousands of discrete strategies, and several security game based systems have been deployed in security practice, see for instance, Tambe and his co-authors' work [14].

Secondly, discrete strategies better reflect reality than continuous strategies. In practise, security performance is not a strictly increasing function of the security investments, instead, it is a stepwise function. The attack performance is analogous. The security or attack performance will not jump to a new level until the increment of the defence investment or attack effort reaches a certain level (e.g., a new scenario can be afforded to be included in the security policy). Hence, security policies are scenario-oriented. The defender firstly evaluates what gaps there are for the designed security level and the current security situation. In the meantime, possible attack scenarios are considered. Subsequently, the defender proposes corresponding defence scenarios. Cost and effectiveness analysis of these proposed scenarios are also carried out. The defender's strategies can therefore be discretely modelled as what kind of defence scenarios to deploy while the attacker's strategies can be discretely modelled as what kind of attack scenario to use. The cost effectiveness analysis procedure provides the information of calculating the payoffs related to the game.

Nonetheless, game theoretic models with continuous strategy sets have their roles in high level security investment problems. For instance, see Nikoofal and Zhuang's work on game theoretically allocating defensive budgets among different cities in the United States [15].

2.2.2 Nash Equilibrium

The so-called 'Nash Equilibrium' (NE) is the most popular solution for non-cooperative games. In a non-cooperative game, players are not able to communicate with each other while their expected payoffs are determined by both their own strategy and other players' strategies. This situation brings a dilemma to players: what would be the most optimal strategy to play, in order to obtain the highest possible payoff?

A straightforward idea is that if one player would know other players' strategies, then he plays the strategy that can bring himself the highest payoff, or, in other words, he plays his best response strategies. For instance, in the game shown in Fig. 2.1 (left-hand side), if the attacker knows that the defender plays a "ND", then the attacker's best response would be playing the strategy "A". However, if the defender would know that the attacker is playing an "A", then the defender's best response would be "D". Therefore, the answer to the dilemma is a group of strategies from each player, and each strategy in the group is a best response to the corresponding player with respect to all other strategies in the group.

A Nash Equilibrium (NE) is a set of players' strategies (one strategy per player), in which each player's strategy in the NE is the best response to all other players' strategies in the NE. For a two-player game, a pure strategy NE (i^*, j^*) satisfies the following condition:

$$U_1(i^*, j^*) \geq U_1(i, j^*), \forall i = 1, 2, \ldots, m \tag{2.3}$$

and

$$U_2(i^*, j^*) \geq U_2(i^*, j), \forall j = 1, 2, \ldots, n \tag{2.4}$$

While a mixed strategy NE (x^*, y^*) satisfies:

$$x^{*T} \cdot U_1 \cdot y^* \geq x^T \cdot U_1 \cdot y^*, \forall x \in X \tag{2.5}$$

and

$$x^{*T} \cdot U_1 \cdot y^* \geq x^{*T} \cdot U_1 \cdot y, \forall y \in Y \tag{2.6}$$

In other words, if players follow a Nash Equilibrium, then they are playing mutual best responses to each other. A NE is thus an action profile a^* with the property that no player i can do better by choosing an action different from a_i^*, given that every other player j adheres to a_j^* [11].

The Nash Equilibrium is named after the American mathematician John Nash, since he proved the existence of the NE in any game with finite players and finite pure strategies (i.e., finite games). Finding a pure strategy NE in a finite game is easy while finding a mixed strategy NE can be quite difficult. Lemke and Howson [4] proposed a linear combinatorial algorithm for finding a mixed strategy NE for finite games with two players.

The NE is a theoretically perfect solution for a simultaneous game, since under the assumption of all players being rational, no player has the motivation to deviate from his/her NE strategy if the opponents do not either. In zero-sum games, the Nash Equilibria are interchangeable, thus there is no NE selection problem in such games,

Fig. 2.2 A simple bi-matrix
game with multiple Nash
Equilibria (NE)

		B	
		F	O
G	F	1,2	0,0
	O	0,0	2,1

see also Proposition 2.1. However, when there is more than one NE in a two-player
non-zero-sum game, it is impossible to predict which NE would be the outcome of
the game. For instance, the battle of the sexes game (BoS), in which there are two
players (a girl and a boy) and the girl prefers to go to the opera (O) while the boy
prefers to play football (F). An illustrative payoff matrix for the BoS is shown in
Fig. 2.2, there are two pure strategy NE (i.e., (F,F) and (O,O)) and 1 mixed strategy
NE (i.e., the girl plays x = (1/3, 2/3) which means that she agrees to play 'F' at
probability 1/3 and to play 'O' at probability 2/3, and the boy plays y = (2/3, 1/3)
which means that he agrees to play 'F' at probability 2/3 and to play 'O' at
probability 1/3. It is obvious that the girl would prefer the (O, O) outcome while
the boy would prefer the (F, F) outcome. In this case, the outcome of this game
cannot be predicted by calculating the NE.

Proposition 2.1 Nash Equilibria in a two-player zero-sum game are interchangeable, that is to say, if mixed strategy pairs (x_1^*, y_1^*) and (x_2^*, y_2^*) are both NEs for a
zero-sum game $G(U, -U)$, then strategy pairs (x_1^*, y_2^*) and (x_2^*, y_1^*) are also NEs for
this game. Furthermore, the player's payoffs for the different equilibria are the same.

Proof (x_1^*, y_1^*) and (x_2^*, y_2^*) satisfy Formulas (2.7), (2.8), (2.9), and (2.10).

$$x_1^* T \cdot U \cdot y_1^* \geq x^T \cdot U \cdot y_1^*, \forall x \in X \tag{2.7}$$
$$x_1^* T \cdot (-U) \cdot y_1^* \geq x_1^* T \cdot (-U) \cdot y, \forall y \in Y \tag{2.8}$$
$$x_2^* T \cdot U \cdot y_2^* \geq x^T \cdot U \cdot y_2^*, \forall x \in X \tag{2.9}$$
$$x_2^* T \cdot (-U) \cdot y_2^* \geq x_2^* T \cdot (-U) \cdot y, \forall y \in Y \tag{2.10}$$

Formula (2.8) indicates $x_1^* T \cdot (-U) \cdot y_1^* \geq x_1^* T \cdot (-U) \cdot y_2^*$. Formula (2.9) indi-
cates $x_2^* T \cdot U \cdot y_2^* \geq x_1^* T \cdot U \cdot y_2^*$.   Formula   (2.10)   indicates
$x_2^* T \cdot (-U) \cdot y_2^* \geq x_2^* T \cdot (-U) \cdot y_1^*$.   Formula   (2.7)   indicates
$x_1^* T \cdot U \cdot y_1^* \geq x_2^* T \cdot U \cdot y_1^*$. Together we have:

$$x_1^* T \cdot U \cdot y_1^* \leq x_1^* T \cdot U \cdot y_2^* \leq x_2^* T \cdot U \cdot y_2^* \leq x_2^* T \cdot U \cdot y_1^* \leq x_1^* T \cdot U \cdot y_1^* \tag{2.11}$$

Therefore, all above inequalities should be equal. Furthermore, we have:

$$x_1^* T \cdot U \cdot y_2^* = x_2^* T \cdot U \cdot y_2^* \geq x^T \cdot U \cdot y_2^*, \forall x \in X \tag{2.12}$$
$$x_1^* T \cdot (-U) \cdot y_2^* = x_1^* T \cdot (-U) \cdot y_1^* \geq x_1^* T \cdot (-U) \cdot y, \forall y \in Y \tag{2.13}$$
$$x_2^* T \cdot U \cdot y_1^* = x_1^* T \cdot U \cdot y_1^* \geq x^T \cdot U \cdot y_1^*, \forall x \in X \tag{2.14}$$

$$x_2^* T \cdot (-U) \cdot y_1^* = x_2^* T \cdot (-U) \cdot y_2^* \geq x_2^* T \cdot (-U) \cdot y, \forall y \in Y \qquad (2.15)$$

Thus, $\left(x_1^*, y_2^*\right)$ and $\left(x_2^*, y_1^*\right)$ are also NEs for game G. Formula (2.11) ensures that players will receive the same payoff for the different equilibria.

Proposition 2.1 is an important property of zero-sum games since in case there are multiple NEs in the game, the player can play any equilibrium. A non-zero-sum game does not have this property, see Fig. 2.2 for instance (the reason is given in the first paragraph of Sect. 2.2.3).

2.2.3   Stackelberg Equilibrium

Simultaneous games may have multiple Nash Equilibria (NE). Therefore, a new dilemma arises, that is, the NE selection problem. One solution to this new dilemma is that some players of the game receive a capability to move first (being the so-called 'game leaders'), and other players follow (being the so-called 'game followers') with knowing the leaders' strategies. For instance, the battle of the sexes game shown in Fig. 2.2 has 3 NEs. Player G prefers the equilibrium (O,O) which brings her a payoff of 2 and in contrast, Player B prefers the equilibrium (F,F) which brings him a payoff of 2. If both players play their preferable NE, i.e., the girl plays strategy "O" and the boy plays strategy "F", then the players are no longer playing a NE and both of them would obtain a payoff of 0. If the girl could move first, for instance, buy two opera tickets for herself as well as for the boy before the battle, then the boy knows that the girl is definitely going to the opera, therefore he will also follow since then his best response is to follow the girl.

A two-player game is called a Stackelberg game if one player moves first and another player follows the leader, knowing the leader's strategy. In the Stackelberg game, the game follower knows the leader's strategy, and therefore is able to play his/her optimal strategy. The leader also knows that the follower knows his/her strategy and the follower will try playing optimally, therefore the leader can also play accordingly.

A so-called 'Strong Stackelberg Equilibrium' (SSE) models the above procedure, and can be formulated as:

$$\bar{y} = \max_{y \in Y} U_l(x(y), y) \qquad (2.16)$$
$$x(y) = \max_{x \in X} U_f(x, y) \qquad (2.17)$$

Formula (2.17) denotes that knowing the leader's committed strategy y, the follower would play his best response strategy $x(y)$. Formula (2.16) indicates that the leader can also work out the $x(y)$, and therefore is able to play optimally. $U_l(x, y)$ and $U_f(x, y)$ denotes the leader and the follower's payoffs in case that the follower plays an "x" and the leader plays a "y".

The equilibrium is called 'Strong Stackelberg Equilibrium' instead of just 'Stackelberg Equilibrium', due to the fact that the so-called "breaking-tie" assumption is involved. The assumption requires that, when the follower has multiple best response strategies to the leader's committed strategy y (we say there is a tie for the follower), the follower would play the best response strategy that maximizes the defender's payoff (we say the follower breaks the tie preferably for the leader). The Stackelberg Equilibrium is not unique as well without the "breaking-tie" assumption.

For instance, for the game shown in Fig. 2.1, if the defender is the game leader and she plays "NotDefend" at probability 1/6 and "Defend" at probability 5/6, then both the "NoAttack" and the "Attack" would be the attacker's best response. In this case, both strategies would bring the attacker a payoff of 0 and on the contrary, the defender's payoff would be $-5/6$ if the attacker responds "NoAttack" and it would be $-8/6$ if the attacker responds "Attack". Under the "breaking-tie" assumption, the attacker would play the "NoAttack" strategy.

However, the "breaking-tie" assumption is anti-intuitive if applied in the security domain. In security, the defender is usually considered as the game leader, and the attacker is the follower. The "breaking-tie" assumption then means that the attacker is playing preferably before the defender! To fix this drawback, von Stengel and Zamir [16] point out that the game leader can deviate a little bit from her SSE strategy, promoting her preferable strategy to be the unique best response to the game follower. For instance, in the above example, the defender may play a mixed strategy $x = (0.99/6, 5.01/6)$ so that the attacker's best response becomes unique, and it is "NoAttack".

2.3 Criticisms on Game Theoretic Models for Security Improvement

A special type of game theoretic models developed for the purpose of improving security, in which what is better for one player is bad for the other player, is defined as Security Game [14]. Security games have been widely studied in academia, and several security game based systems have been deployed in reality [14]. However, criticisms do exist.

Some security game models are criticized as 'magic mathematical games' due to sometimes unrealistic assumptions. Most researchers agree that (human) adversaries would plan and implement attacks adaptively. However, whether adversaries are rational (i.e., aiming at maximizing their payoff) is still a topic under study. Researchers also realize that security risk management involves huge uncertainties such that the 'common knowledge assumption' would not hold. For a more detailed discussion of these criticisms, interested readers are referred to Guikema [17].

Besides its possible unrealistic assumptions, game theoretic modelling is also criticized for its requirements with respect to quantitative input. As illustrated in Fig. 2.1, parameters such as the defender's defence cost (C_d), the defender's loss from a successful attack (L), the attacker's attack cost (C_a), the attacker's penalty (P)

from a failed attack, and the attacker's reward from a successful attack (R) should be provided in order to analyse the game. In practice, however, it can be quite difficult (almost impossible) to obtain these exact data. Let us take as an example R, which denotes the attacker's gain from successfully attacking the target: it is not possible, in practise, to know what would be the exact gain for the attacker, since it is largely dependable on the attacker's perception. In literature, the Chemical Plant Protection game proposed by Zhang and Reniers [18] requires quantitative data such as the success probabilities and consequences of an attack under any given attack scenarios and any given defence plans, from both the defender and the attacker's point of view. In the work of Feng et al. [19], the defender needs to know a prior probabilities of occurrence of different types of attackers, and also attackers' estimations of vulnerabilities and consequences under each of the players' strategy pairs. These above mentioned quantitative inputs are very difficult to obtain.

2.4   Integrating Conventional Security Risk Assessment Methodologies and Game Theory for Improving Chemical Plant Protection

Conventional security risk assessment methodologies, such as the SRFT and the API SRA methodology, being developed by chemical security experts and practitioners, are systematic and practically implementable for security risk assessment in the process and petrochemical industries. These methodologies, since released, have been extensively used in industrial practice and have been much referred to in academic research. As also mentioned in Chap. 1, a common drawback of these methodologies is their failure on modelling the intelligent interactions between the malicious attackers and the security defender.

Game theory was created to deal with intelligent interactions among multiple strategic actors, while its drawback on the application in the security domain is that it is too far away from the security practise. Security experts and practitioners are not familiar with game terminologies and they do not like the complex mathematical formulas in the development procedure of the games.

Zhang et al. [20] proposed an approach for integrating the conventional security risk methods in the chemical security domain and game theoretic models, as shown in Fig. 2.3. In their approach, conventional security methods (e.g., the API SRA methodology) act as a bridge between the industrial practise and the game modellers. At the first step, the conventional security method should be employed, for screening the chemical plant, identifying critical assets, evaluating threats/vulnerabilities/consequences etc. After the first step, a certain set of security related data will be obtained, for instance, see Fig. 1.6 for the output data of the API SRA method. At the second step, instead of analysing the output data from the first step by using risk matrices or by using risk ranking systems (as used in conventional security methods), game theoretic analysis are introduced to deal with these data (output

Fig. 2.3 A framework of integrating the API SRA methodology and game theory

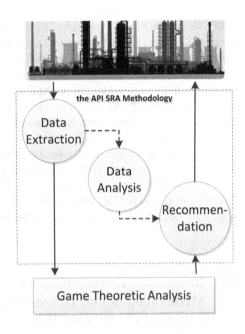

from the first step). At the third step, after analysing the data in a game theoretic approach, we must not show the game theoretic results in game terminologies to security experts and practitioners directly. Alternatively, the game theoretic results should be translated back to the conventional security risk assessment terminologies, for instance, by reflecting the attacker's mixed equilibrium strategy (in game theory terminology) to the target's attractiveness to the attacker (in API SRA terminology).

2.5 Conclusion

In this chapter, game theory was briefly introduced. Players, strategies, and payoffs are the three components of a game theoretic model, while the 'common knowledge assumption' and the 'rationality' assumption are the two most frequently mentioned assumptions in game theoretic research. Furthermore, the simultaneous moving game and the sequential moving game are explained.

Game theoretic models developed for the purpose of improving security are defined as security games, and these games mostly are two-player (defender and attacker) games with discrete strategy sets. A discussion of the advantages and disadvantages of games with continues strategy sets and of games with discrete strategy sets are given. The Nash Equilibrium for a two-player simultaneous game and the Stackelberg Equilibrium for a leader-follower sequential game are defined.

A discussion on the drawbacks of game theory if applied in the chemical security domain, is given. Furthermore, we propose an approach on integrating conventional security risk assessment methods and game theory for the purpose of improving security in the chemical and process industries.

References

1. Von Neumann J, Morgenstern O. Theory of games and economic behavior. Princeton: Princeton University Press; 2007.
2. Nash JF. Equilibrium points in n-person games. Proc Nat Acad Sci USA. 1950;36(1):48–9.
3. Harsanyi JC. Games with incomplete information played by "Bayesian" players, i–iii: part i. the basic model. Manag Sci. 2004;50((12_suppl)):1804–17.
4. Lemke CE, Howson J, Joseph T. Equilibrium points of bimatrix games. J Soc Ind Appl Math. 1964;12(2):413–23.
5. Chen X, Deng X, editors. Settling the complexity of two-player Nash equilibrium. Foundations of Computer Science, 2006 FOCS'06 47th Annual IEEE Symposium on; 2006: IEEE.
6. Nisan N, Roughgarden T, Tardos E, Vazirani VV. Algorithmic game theory. Cambridge: Cambridge University Press; 2007.
7. 101 GT. The support of mixed strategies. Available from: http://gametheory101.com/courses/game-theory-101/support-of-mixed-strategies/
8. Rios J, Insua DR. Adversarial risk analysis for counterterrorism modeling. Risk Anal. 2012;32 (5):894–915.
9. Kelly A. Decision making using game theory: an introduction for managers. Cambridge: Cambridge University Press; 2003.
10. Friedman M. Essays in positive economics. Chicago: University of Chicago Press; 1953.
11. Osborne MJ. An introduction to game theory. New York: Oxford University Press; 2004.
12. Pita J, Jain M, Tambe M, Ordóñez F, Kraus S. Robust solutions to Stackelberg games: addressing bounded rationality and limited observations in human cognition. Artif Intell. 2010;174(15):1142–71.
13. Skaperdas S. Contest success functions. Econ Theory. 1996;7(2):283–90.
14. Tambe M. Security and game theory: algorithms, deployed systems, lessons learned. New York: Cambridge University Press; 2011.
15. Nikoofal ME, Zhuang J. Robust allocation of a defensive budget considering an attacker's private information. Risk Anal. 2012;32(5):930–43.
16. Von Stengel B, Zamir S. Leadership with commitment to mixed strategies. In: CDAM Research Report LSECDAM-2004-01. London School of Economics; 2004.
17. Guikema SD. Game theory models of intelligent actors in reliability analysis: an overview of the state of the art. Game theoretic risk analysis of security threats. New York: Springer; 2009. p. 13–31.
18. Zhang R. A game-theoretical model to improve process plant protection from terrorist attacks. Risk Anal. 2016;36(12):2285–97.
19. Feng Q, Cai H, Chen Z, Zhao X, Chen Y. Using game theory to optimize allocation of defensive resources to protect multiple chemical facilities in a city against terrorist attacks. J Loss Prev Process Ind. 2016;43:614–28.
20. Zhang L, Reniers G, Chen B, Qiu X. Integrating the API SRA methodology and game theory for improving chemical plant protection. J Loss Prev Process Ind. 2018;51(Suppl C):8–16.

Chapter 3
Single Plant Protection: A Game-Theoretical Model for Improving Chemical Plant Protection

In this chapter, we introduce a game theoretic model for protecting a chemical plant from intelligent attackers. The model is named Chemical Plant Protection Game, abbreviated as "CPP Game" [1]. The CPP Game is developed based on the general intrusion detection approach in chemical plants. To this end, the general intrusion detection approach is firstly introduced. We develop and explain the CPP Game by modelling its players, strategies, and payoffs. Afterwards in Sect. 3.3, different equilibrium concepts are used to predict the outcome of the CPP Game [2]. An analysis of the inputs and outputs of the game is provided in Sect. 3.4, from an industrial practice point of view [3]. Finally, conclusions are drawn at the end of this chapter.

3.1 General Intrusion Detection Approach in Chemical Plants

Though security risk assessment is a relatively new topic stimulated by the 9/11 attack, chemical industries have a long history with respect to separating their assets and facilities from citizens and nearby communities. The main purpose of this isolation is the existence of large amounts of dangerous materials, which, in case of a major accident, might lead to losses suffered by the plant as well as by the surrounding communities. The separation is achieved by distance (using perimeters), evidently, but also by intrusion control.

Figure 3.1 shows a typical illustrative layout of a chemical plant, indicating its general physical intrusion detection approach. As shown in the figure, the terms "PERIMETER", "ENTRANCE", and "ZONE" are used. A perimeter concerns the

This Chapter is mainly based on three published papers: Zhang and Reniers [1, 2], Zhang et al. [3].

© Springer International Publishing AG, part of Springer Nature 2018
L. Zhang, G. Reniers, *Game Theory for Managing Security in Chemical Industrial Areas*, Advanced Sciences and Technologies for Security Applications, https://doi.org/10.1007/978-3-319-92618-6_3

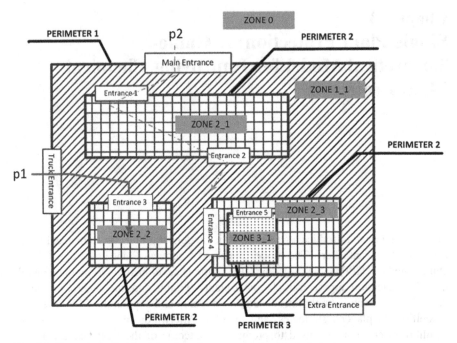

Fig. 3.1 General physical intrusion detection approach in chemical plants

boundary of an area related to the plant, and may consist of a fence, a wall, or even a geographical boundary such as a river bank. Whatever it is composed of, a perimeter is a closed area, and it should prevent illegal intrusion. Entrances are attached to perimeters. Authorized people are checked at entrances and afterwards they are allowed to pass the entrance. Zones are generated automatically due to the perimeters. As shown in Fig. 3.1, zones are distinguished by different levels and each level may have several sub-zones. For instance, zone level 2 in Fig. 3.1 has 3 sub-zones. Moreover, a higher level zone (e.g., level i) is contained in a lower level zone (i.e., the level $i − 1$). Due to the possibility of being detected in each entrance and in each zone, an intruder could evidently easier reach a target situated in lower level zones. To this end, important infrastructures (based on for instance sensitivity, confidentiality, dangerousness, and what have you) of the chemical plant are usually situated in higher level zones.

With the general intrusion detection system, the industrial manager may allocate its security resources at each entrance or in each zone. For instance, the main entrance of a plant could be equipped with an employee card recognition machine, a security guard, a communication system (to the local police station or to the security centre of the plant), while a camera system (e.g. CCTV) and regular guard patrolling forms a typical detection scenario in zones. It is worth noting that security scenarios at each entrance or zone are not necessarily fixed, that is, the defender may have several different scenarios at one place, and she chooses a scenario according to the threat level. Different scenarios represent different security alert levels (SALs). If the plant evaluates that it has a higher threat level, either based on intelligence gathered or based on past security events, the plant will increase its SAL at some entrances or zones. Higher SALs need enhanced security scenarios. For instance, in case of a lower SAL, the security guards are not armed, while in a higher SAL, they could be armed (depending on the world region where the plant is located). In a higher SAL, the intruder will be more likely to be detected and halted, but the cost of a higher SAL is usually higher.

An intruder would have to, firstly, choose a target to attack, secondly, choose an attack scenario, and thirdly, choose a critical (easiest for him) path to reach the target. It is a complex decision problem, since there are multiple facilities and areas inside a plant, and the attacker would choose the target according to his purpose. A terrorist, for example, would prefer to cause some leakage or explosion at some storage tank (s), production facilities etc., while an environmental activist, might prefer to shut down the power station of a plant to stop the operation of the plant. Different types of attackers will use different attack scenarios, to different targets. The scenario of a terrorist attack with respect to a toxic tank can be related to the use of an explosive device, while the scenario with respect to a production facility can be linked to the switching off of an important safety valve. The attacker also needs to choose an intrusion path to reach the target. An intrusion path consists of several entrances where each entrance belongs to a different level of perimeter. For instance, the intrusion path p1 in Fig. 3.1 consists of the truck entrance and entrance 3. Furthermore, in order to simplify the research and to reflect reality, we assume that the intruder would never step into the same zone level twice. This assumption excludes paths such as p2 in Fig. 3.1, since if p2 is followed, the intruder would step into zone level 1 twice (after the main entrance and after entrance 2). This assumption is useful for simplifying the path analysis. Otherwise without this assumption, the intruder may have infinite numbers of intrusion paths. This assumption also reflects reality, since if the intruder steps into the same zone level twice, he would have to pass more entrances and zones, which will increase the likelihood of being detected.

Figure 3.2 is a plot from the intruder's viewpoint, illustrating the intrusion and attack procedure for the case of a chemical plant, as shown in Fig. 3.1. The indexes of zones and perimeters are using the same name/label as used in Fig. 3.1, while entrance A_{rj} denotes the jth access of the perimeter r. Without loss of generality, we can map A_{11} in Fig. 3.2 to the main entrance in Fig. 3.1; A_{12} to the truck entrance; and so forth. Besides intruding through an entrance, the intruder may also step over the perimeter, which is also considered in Fig. 3.2 as an access. The red dot line in Fig. 3.2 represents the intrusion path p1 in Fig. 3.1. As discussed earlier, the defender allocates security resources at entrances and in zones, while the intruder would have to pass the entrances and zones being situated on his intrusion path. We define a $P_i^z \in [0, 1]$, denoting the probability of successfully passing zone level i, and we define a $P_r^p \in [0, 1]$, denoting the probability of passing the perimeter r. Both P_i^z and P_r^p should be determined by the defender's security alert level at the entrance or the zone, and by the attacker's attack scenario. A more detailed discussion of P_i^z and P_r^p

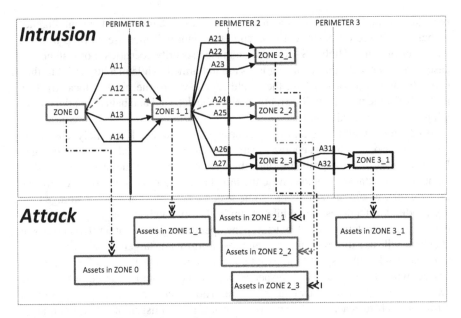

Fig. 3.2 The intrusion and attack procedure

will be given in Sect. 3.2.3. In theoretical research, the contest success function (CSF) [4] is often used to estimate these two parameters.

Since there might be multiple entrances on one perimeter (which is always the case in industrial practice), the probability of successfully passing the perimeter P_r^p will be equal to the probability of successfully passing the chosen entrance j, denoted as P_{rj}^p. For instance, in the intrusion path p1, we would have $P_1^p = P_{12}^p, P_2^p = P_{24}^p$ (see Fig. 3.2). The security alert levels at different entrances of a perimeter can be different, thus the intruder has different success probabilities by choosing different entrances. However, it is not necessary that the intruder always chooses the easiest entrance to intrude. This is the result of the detection in the zones: an easier passing entrance might be situated further (in distance) to the target than a more difficult passing entrance, thus although the intruder could easily pass the entrance, the probability of him being detected in the zone will be higher.

Based on above analysis, if the intruder aims to attack a target situated in zone level 1, the probability that he will successfully reach the target can be calculated as shown in Formula (3.1).

$$P = \prod_{i=0}^{I} P_i^z \cdot \prod_{r=1}^{I} P_r^p. \qquad (3.1)$$

3.2 Game-Theoretical Modelling: The Chemical Plant Protection Game (CPP Game)

We introduce and explain the chemical plant protection game (CPP Game) from three modelling perspectives, namely, that of the players (Sect. 3.2.1), the strategies (Sect. 3.2.2), and the payoffs (Sect. 3.2.3).

3.2.1 Players

The chemical plant protection game is played between two players: the defender and the attacker.

The 'defender' represents the security department of a chemical facility, who is responsible for the security management (amongst others carrying out security risk assessments) of the plant. The attacker can be various, for instance, terrorists, criminals, and environmental activists. The game is modelled as a two-player game implying that the collusion among different types of attackers are excluded from this research. In reality, different types of attackers may cooperate to implement an attack. A very straightforward cooperation can be, for instance, a disgruntled employee working together with criminals to cause damages and losses to the plant. If collaborative partnerships among attackers are taken into consideration, the model will be a multiple players game, and it will involve both a cooperative game (among different attackers) and a non-cooperative game (between the defender and the attackers). For legibility and simplicity reasons, in this research, we ignore the case of an alliance of attackers, and assume that different attackers are independent to each other.

The existence of different types of attackers can be modelled in a Bayesian approach. In a chemical plant protection game where multiple types of attackers are considered, a prior probability can be assigned to each type of attacker. Moreover, these prior probabilities can be calculated based on the threat level information, as shown in Formula (3.2). With the prior probability, the Bayesian chemical plant protection game can be interpreted as a defender-attacker two-player game in which the defender is facing threat t with probability p^t. By employing a Bayesian approach, not only the collusion among different types of attackers is not modelled, but also the simultaneous occurrence of more than two types of attacker is ignored. The probability p of a threat t can be expressed as:

$$p^t = \frac{ts^t}{\sum_{t \in TL} ts^t}, t \in TL, ts \in TS. \tag{3.2}$$

In which p^t denotes the prior probability of threat t, ts^t represents the threat level of threat t, TL and TS are the threat list and the threat level set respectively. This threat information can be obtained by some conventional security risk assessment results, using the API SRA [5], for instance. Section 3.4.1 shows more details about obtaining the inputs for a chemical plant protection game.

The assumption of rationality can be justified for the defender, being the security department of a chemical plant. However, assuming rationality for the attacker is not so straightforward and requires more explanation and interpretation. As mentioned in the previous chapter, rationality in game theory is defined as the maximisation of a player's own payoff. Thus although from an ordinary person's point of view, some attacker behaviours (e.g., terrorist suicide attacks) are very emotion-based, intelligent attackers do have their own goal and they plan their attack to maximize, for instance, the defender's damage (in case of terrorism). Therefore, if the attacker's payoff is modelled according to the attack goal, then the rationality assumption can be defended. In Chap. 5, we extend the chemical plant protection game to deal with bounded rational attackers.

Another interesting topic about modelling from the viewpoint of the player concerns whether the players have complete information of the game or not. Research has pointed out that terrorists need to collect certain information before they implement an attack. In fact, some research reveals that terrorists are able to obtain at least 80% (in some cases even 100%) of the needed information through public access [6]. According to this research, it is reasonable to assume that the attackers have complete information of the game. However, due to the lack of historic data, the defender hardly has any information about the attackers, and the required amount of information that the defender needs to dispose of, might thus be scarce and very hard to obtain. For instance, the attacker can easily estimate the defender's defence costs for a security scenario (e.g., a combination of camera and patrolling), while it can be quite difficult for the defender to know the attacker's cost of obtaining an explosive device and/or to know the attacker's total available budget. To meet this modelling challenge, in Chap. 4, we propose a chemical plant protection game in which the defender only knows a distribution-free interval for the attacker's parameters. In the basic chemical plant protection game (in Chap. 3), we assume that both the defender and the attacker have complete information of the game.

In conclusion, the chemical plant protection game is a two-player game played by a defender and an attacker, and the types of attacker can be various. Both players have complete information of the game and both are assumed to behave rationally. The information related to the game is common knowledge to both players.

3.2.2   Strategies

To secure a chemical plant, the defender may set different security alert levels (SAL) at each entrance and in each zone of the plant. Table 3.1 demonstrates an example of corresponding countermeasures with different security alert levels. A pure strategy of the defender can be defined as a combination of security alert levels at each entrance and in each (sub-) zone, as shown in Formula (3.3):

Table 3.1 Illustrative countermeasures corresponding to different security alert levels

SAL	Corresponding measures	
	Entrances	(Sub-) Zones
1 (White)	Access Control: wearing badge	Light: minimum perimeter
2 (Yellow)	Access Control: badge reader	Light: all the zone, for sight
	CCTV: security check	
3 (Orange)	Access Control: badge reader, restricted list	Light: all the zone, for video
	CCTV: security check, cyclical check at perimeter	Dog detection: dog nearby
4 (Red)	Access control: badge reader, explosive search	Light: all the zone and towards outside, for video
	CCTV: security check, permanent check at perimeter	Dog detection: dog with microphone inside

$$s_{di} = z^0 \times \prod_{r=1}^{Q} \left(A_1^r \times A_2^r \times \ldots \times A_{ent(r)}^r \times z_1^r \times z_2^r \times \ldots \times z_{sub(r)}^r \right) \qquad (3.3)$$

In which z^0 denotes the SAL in zone level 0 (i.e., the outside zone); Q represents the total zone levels in the plant (for instance, the plant shown in Fig. 3.1 has a $Q = 3$); A_j^r is the SAL at the jth entrance of perimeter r; $ent(r)$ is the number of entrances of perimeter r (for instance, the plant in Fig. 3.1 we have $ent(1) = 4$, ent $(2) = 7$, $ent(3) = 2$); z_i^r denotes the SAL in the ith sub-zone of zone level r; $sub(r)$ denotes the number of sub-zones in zone level r (for instance, in Fig. 3.1, we have $sub(1) = 1$, $sub(2) = 3$, $sub(3) = 1$); $\times$ and $\prod$ represent the Cartesian product.

For simplicity reasons, we assume that the defender has the same number of SALs at each entrance point and in each subzone, say, k. For instance, a plant may set the security alert level at its main entrance as low/medium/high, thus we have $k = 3$. In other cases the security alert level can be white/blue/yellow/orange/red/, thus we have $k = 5$. The total number of the defender's pure strategies n can therefore be calculated by Formula (3.4):

$$n = k^{1+\sum_{r=1}^{Q}(ent(r)+sub(r))} \qquad (3.4)$$

We further define the defender's pure strategy set as $S_d = \{s_{d1}, s_{d2}, \ldots, s_{dn}\}$.

An attacker's pure strategy is modelled as the combination of (i) which target to attack; (ii) with what attack scenario; and (iii) from which intrusion path to reach the target, as formulated in Formula (3.5):

$$s_{ai} = target \times \prod_{r=1}^{I} j_r \times e \qquad (3.5)$$

In which *target* denotes the facility that the attacker is going to attack; I represents the zone level that the target is situated in; e is the attack scenario, or the attack effort;

j_r is the entrance that the attacker chooses to pass perimeter r, and $j_r \in \{1, 2, \ldots, ent(r)\}$.

An example of the attacker's pure strategy is the following: if the attacker follows intrusion path p1 in Fig. 3.1, and he wants to attack a target in zone 2_2 (assume the target has an index $\mathcal{L}$) with an explosive device, say a bomb, then this pure strategy can be expressed as: $s_{ai} = \mathcal{L} \times \textit{Truck Ent} \times \textit{Ent3} \times \textit{Bomb}$.

The total number of pure strategies of the attacker m can be calculated by Formula (3.6), and we define the attacker's pure strategy set as: $S_a = \{s_{a1}, s_{a2}, \ldots, s_{am}\}$. In Formula (3.6), k_a^{tgt} is the number of attack scenarios that the attacker may use to the target tgt; Ast_0 denotes the set of targets in zone 0; $Ast_{r,i}$ denotes the set of targets that are situated in the ith subzone in zone level r; $ent(j, i)$ represents the number of entrances on perimeter j that can lead to sub zone i. For instance, in Fig. 3.1, we have $ent(2) = 7$ and $ent(2, 2\_1) = 3$. It is worth noting that the defender can hardly enumerate all the available attack scenarios, thus the k_a can be difficult to know [7]. In the chemical plant protection game, we simply list all the possible attack scenarios according to the defender's knowledge and experiences, which aims to make the best use of the available data/knowledge.

$$m = \sum_{tgt \in Ast_0} k_a^{tgt} + \sum_{r=1}^{Q} \sum_{i=1}^{sub(r)} \left(\sum_{tgt \in Ast_{r,i}} k_a^{tgt} \cdot \prod_{j=1}^{r} ent(j, i) \right) \quad (3.6)$$

According to Formulas (3.4, 3.5 and 3.6), we notice that both the defender and the attacker will have a finite number of pure strategies (i.e., n and m respectively). The number of pure strategies of the defender will increase dramatically as the scale of the plant grows. However, this is not a problem in industrial practice, as the defender's pure strategy set can be cut according to her available budget, see, for instance, Talarico et al. [8].

The players' mixed strategy space can be defined as $Y = \{y \in R^{|S_d|} | \sum y_i = 1, y_i \in [0, 1]\}$, and $X = \{x \in R^{|S_a|} | \sum x_i = 1, x_i \in [0, 1]\}$, for the defender and for the attacker, respectively.

3.2.3 Payoffs

Payoffs are numbers representing the players' motivations. In the chemical plant protection game, the defender's payoff is defined as her expected loss from an attack minus her defence cost, while the attacker's payoff is defined as his expected gain from an attack minus his attack cost, as formulated in Formulas (3.7) and (3.8) respectively.

$$u_d(s_a, s_d) = -\left(P(s_a, s_d) \cdot P_y(s_a) \cdot L_y(s_a) + C_d(s_d)\right) \quad (3.7)$$

$$u_a(s_a, s_d) = \tilde{P}(s_a, s_d) \cdot \widehat{P_y}(s_a) \cdot \tilde{L}_y(s_a) - C_a(s_a) \quad (3.8)$$

In which: (s_a, s_d) denotes a given attacker and defender pure strategy pair; P is the probability that the attacker will successfully reach the target, and it is calculated by Formula (3.1); P_y is the probability that the attacker will successfully implement the attack under the condition that he has reached the target; L_y is the estimated loss assuming that the attack is successfully implemented; $\tilde{P}$, $\widehat{P_y}$, and $\tilde{L}_y$ have the same meaning as P, P_y, and L_y, but they are the parameters estimated from the attacker's point of view; C_d and C_a are the defence and attack cost, respectively.

It is worth noting that for one and the same parameter, the defender and the attacker may have different perceptions and hence different estimations. For instance, for a thief, stealing a computer from a control room would be his true goal and would lead to obtaining the hardware, but the defender may be most interested in the potential loss of important data (e.g., technique documents). Thus in this case, we may have $L_y \gg \tilde{L}_y$. Furthermore, for probabilities of successfully going through some entrances or zones, the defender and the attacker can also have quite different estimations: for a risk-seeking intruder, we may have $P \leq \tilde{P}$; for a risk-averse intruder, we may have $P \geq \tilde{P}$.

By implementing Formulas (3.7) and (3.8) for each defender and attacker pure strategy pair, we will obtain their payoff matrices, denoted as U_d and U_a respectively. According to the players' pure strategy numbers, we know that both payoff matrices are $m \times n$ matrices. If there are multiple types of attackers, we denote the defender and the attacker's payoff as U_d^t, U_a^t, for $\forall t \in TL$, respectively.

Formulas (3.7) and (3.8) also reveal that the chemical plant protection game is not necessarily a zero-sum game, though defenders and attackers always have opposite interests. The non-zero-sum property of the CPP game contains two aspects: (i) both the defence cost C_d and the attack cost C_a are involved in the payoff definitions, and no player will benefit from the other player's behaviour cost; (ii) the defender and the attacker might evaluate the same parameters with different values, including probabilities (e.g., P and $\tilde{P}$) and consequences (i.e., L and $\tilde{L}$).

However, in some special conditions, the CPP game can be a strategically zero-sum game, as explained hereafter. Re-write the payoff Formulas (3.7) and (3.8) as Formulas (3.9) and (3.10).

$$u_d(s_a, s_d) = -f(s_a, s_d) - C_a(s_a) \tag{3.9}$$
$$u_a(s_a, s_d) = \tilde{f}(s_a, s_d) - C_d(s_d) \tag{3.10}$$

In which:

$$f(s_a, s_d) = P(s_a, s_d) \cdot P_y(s_a) \cdot L_y(s_a) - C_a(s_a) + C_d(s_d) \tag{3.11}$$
$$\tilde{f}(s_a, s_d) = \tilde{P}(s_a, s_d) \cdot \widehat{P_y}(s_a) \cdot \tilde{L}_y(s_a) - C_a(s_a) + C_d(s_d) \tag{3.12}$$

Define a zero-sum defender-attacker game $(F, -F)$, of which the defender and the attacker's strategy sets are the same as the defender and the attacker's strategy sets in the CPP game, and payoff units of the new game are defined by Formulas (3.11) and (3.12).

If, in some cases, for all strategy tuples of the CPP game, the condition $\tilde{P} \cdot \tilde{P}_y \cdot \tilde{L} = P \cdot P_y \cdot L$ holds, then $(\bar{x}, \bar{y})$ is a NE of the CPP game if and only if $(\bar{x}, \bar{y})$ is a NE of the game $(F, -F)$.

Proof Since $\tilde{P} \cdot \tilde{P}_y \cdot \tilde{L} = P \cdot P_y \cdot L$, we directly know that $\tilde{f} = f$. According to the definition of NE, $(\bar{x}, \bar{y})$ is a NE of the CPP game $\Leftrightarrow$

$$\begin{cases} \bar{x}^T \cdot U_a \cdot \bar{y} \geq x^T \cdot U_a \cdot \bar{y}, & \forall x \in X \\ \bar{x}^T \cdot U_d \cdot \bar{y} \geq \bar{x}^T \cdot U_d \cdot y, & \forall y \in Y \end{cases}$$

$$\Leftrightarrow \begin{cases} \bar{x}^T \cdot F \cdot \bar{y} - \bar{x}^T \cdot C_D \cdot \bar{y} \geq x^T \cdot F \cdot \bar{y} - x^T \cdot C_D \cdot \bar{y}, & \forall x \in X \\ -\bar{x}^T \cdot F \cdot \bar{y} - \bar{x}^T \cdot C_A \cdot \bar{y} \geq -\bar{x}^T \cdot F \cdot y - \bar{x}^T \cdot C_A \cdot y, & \forall y \in Y \end{cases} \qquad (3.13)$$

In the above formulas, C_D and C_A are the defence and the attack cost matrices respectively, Their entries at the ith row (for C_D), j^{th} column (for C_A) are the defender's defence cost and the attacker's attack cost respectively, when the attacker plays a pure strategy s_i^a and the defender plays s_j^d. Since the attacker's strategy doesn't influence the defender's defence cost, C_D shows identical rows; and analogously, C_A has identical columns. Therefore, we have $\bar{x}^T \cdot C_D \cdot \bar{y} = \sum_{j \in N} C_{dj} \cdot \bar{y}_j = x^T \cdot C_D \cdot \bar{y}$, and $\bar{x}^T \cdot C_A \cdot \bar{y} = \sum_{i \in M} \bar{x}_i \cdot C_{ai} = \bar{x}^T \cdot C_A \cdot y$. Thus Formula (3.13) becomes:

$$\begin{cases} \bar{x}^T \cdot F \cdot \bar{y} \geq x^T \cdot F \cdot \bar{y}, & \forall x \in X \\ -\bar{x}^T \cdot F \cdot \bar{y} \geq -\bar{x}^T \cdot F \cdot y, & \forall y \in Y \end{cases} \qquad (3.14)$$

Formula (3.14) represents that $(\bar{x}, \bar{y})$ is a NE of game $(F, -F)$.

The proof of the above observation implies that under the condition that $\tilde{P} \cdot \tilde{P}_y \cdot \tilde{L} = P \cdot P_y \cdot L$ holds for all strategy tuples, the CPP game is a strategically zero-sum game [9]. In this case, the analysis of the CPP game becomes easier. For more information of the strategically zero-sum game, interested readers are referred to Moulin and Vial [9].

Although the condition of $\tilde{P} \cdot \tilde{P}_y \cdot \tilde{L} = P \cdot P_y \cdot L$ is a strong condition for the CPP game, it might be the case in some real industrial practice situations. For instance, if the defender and the attacker evaluate the intrusion probabilities, the consequences of an attack etc. in the same way, we would have $\tilde{P} = P$, $\tilde{P}_y = P_y$, and $\tilde{L} = L$, and then the condition holds definitely.

3.3   Solutions for the CPP Game

A solution of a game is a pair of (mixed) strategies that the players would play. In this section, the Nash equilibrium (NE), the Stackelberg equilibrium (SE), the Bayesian Nash equilibrium (BNE), and the Bayesian Stackelberg equilibrium

(BSE) are used to solve the chemical plant protection game. If the defender and the attacker in the game move simultaneously, an NE must be used, while if they move sequentially (defender moves first and attacker follows), a SE must be used. If only one type of attacker is considered, then a NE or a SE should be used, and if multiple types of attackers are considered, a BNE or a BSE should be used.

3.3.1 Nash Equilibrium

In the CPP game, the defender implements her daily defence plan by setting security alert levels at each entrance or in each zone. The CPP game is played simultaneously in the case that when the attacker implements his attack, he does not know any information about the defender's defence. A Nash equilibrium can be employed to predict the outcome of a simultaneous CPP game.

A pure strategy Nash equilibrium (s_a^*, s_d^*) for the CPP game satisfies the condition in Formulas (3.15) and (3.16).

$$u_a(s_a^*, s_d^*) \geq u_a(s_a, s_d^*), \quad \forall s_a \in S_a \tag{3.15}$$

and

$$u_d(s_a^*, s_d^*) \geq u_d(s_a^*, s_d), \quad \forall s_d \in S_d \tag{3.16}$$

A mixed strategy Nash equilibrium (x^*, y^*) for the CPP game satisfies the condition in Formulas (3.17) and (3.18).

$$x^{*T} \cdot U_a \cdot y^* \geq x^T \cdot U_a \cdot y^*, \quad \forall x \in X \tag{3.17}$$

and

$$x^{*T} \cdot U_d \cdot y^* \geq x^{*T} \cdot U_d \cdot y, \quad \forall y \in Y \tag{3.18}$$

The CPP game is a finite game (with two players and each player has a finite number of pure strategies), thus at least one Nash equilibrium exists. However, in most cases, there is no pure strategy Nash equilibrium for the CPP game, since generally (not always, due to the existence of the defence cost and the attack cost) the defender and the attacker have opposite interests.

For calculating a NE for the CPP game, the Lemke and Howson algorithm [10] can be employed.

3.3.2 Stackelberg Equilibrium

If the attacker would know the defender's defence plan when he attacks, the CPP game is a sequential game. The attacker can know the defender's plan by collecting information or by continued and thorough study and observation. Although currently some game-theoretic models in the security domain propose the use of simultaneous games (e.g. [11]), three reasons enforce modellers to prefer modelling the security game as a sequential game.

Firstly, a sequential game can reflect reality better. Literature has shown that adversaries may collect 80% (sometimes even 100%) of the needed information to execute a successful attack [6]. Based on such evidence, we can assume that the attacker has complete information of the security game. However, in industrial practice regarding critical infrastructure protection, the defender usually has to implement her countermeasures (strategy) first, and when the attacker plans the attack, he not only is able to collect the information of the target, but also information of the defender's defence strategies. Thus we may assume that the attacker has both complete and perfect information of the (sequential) game.

Secondly, a sequential game might bring higher payoff to the leader (which is the first-mover, hence the defender) than the simultaneous game. This principle is called "First-mover Advantage" [12]. As was already mentioned, the attacker would also collect information on the defender's strategies. If the attacker can fully observe the defender's executed strategy, the game is a "perfect information game", or a sequential move game; if he cannot observe the defender's strategy, the game is called an "imperfect information game", or a simultaneous game (see also Chap. 2); if he can partly observe the defender's strategy, the case becomes more complicated. The problem is that, the defender, who moves first, does not know whether the attacker can fully observe her strategy or not, thus she does not know whether she is playing a simultaneous game or a sequential game or even a more complicated game. In these cases, Zhuang and Vicki [12] proved that if the attacker's best-response set is a singleton, then the defender's gain from the sequential game is at least the same as that from the simultaneous game (hence the existence of the principle of "First-mover Advantage"). Knowing the "First-mover Advantage", the defender could choose to make her strategy public, to enforce the game being a sequential game.

Thirdly, the equilibria selection problem can be avoided by playing a sequential game. The Nash Equilibrium (NE) [13] is the most extensively used concept in a simultaneous game to predict the outcome. However, as shown in Sect. 2.2.2, a non-zero-sum game becomes unpredictable when there are multiple NEs. In the game illustrated in Fig. 2.2, if the girl would move first, she commits to the boy that she will play the strategy 'O', enforcing the boy to also play 'O', thus the game becomes predictable. Since the security game is not necessarily zero-sum, and it is possible to have multiple NEs, playing the game sequentially can make the game predictable and controllable for the defender.

In a sequential move CPP game, the Stackelberg equilibrium can be employed. Formulas (3.19) and (3.20) define the Strong Stackelberg equilibrium $(\bar{s}_a, \bar{y})$ for the CPP game.

$$\bar{y} = \text{argmax}_{y \in Y} U_d(\bar{s}_a, :) \cdot y \qquad (3.19)$$

$$\bar{s}_a = \text{argmax}_{s_a \in S_a} U_a(s_a, :) \cdot y \qquad (3.20)$$

Formula (3.20) denotes the fact that knowing the defender's mixed strategy y, the attacker will choose a best response pure strategy (i.e., $\bar{s}_a$). Formula (3.19) shows that the defender can also work out the attacker's best response, and thus she can play accordingly.

It is worth noting that in Formula (3.19), the defender can play a mixed strategy, which means that the attacker is only able to know the defender's defence plan, without knowing the defender's exact defence when he attacks. For instance, after a long time observation or by getting the plant's security schedule document, the attacker could know that the plant will set the SAL at the main entrance at low (high) level with a probability of 60% (40%). However, the attacker is assumed not to know whether the defender sets the SAL at the main entrance as low or as high, at the day that he attacks. This is a reasonable assumption since preparing for an attack needs time. Nonetheless, if the attacker knows the exact defence when he attacks, a pure strategy Stackelberg equilibrium $(\bar{s}_a, \bar{s}_d)$ should be employed, as defined in Formulas (3.21) and (3.22).

$$\bar{s}_d = \text{argmax}_{s_d \in S_d} U_d(\bar{s}_a, s_d) \qquad (3.21)$$

$$\bar{s}_a = \text{argmax}_{s_a \in S_a} U_a(s_a, s_d) \qquad (3.22)$$

For calculating the strong Stackelberg equilibrium for the CPP game, the MultiLPs [14] algorithm can be employed.

3.3.3 Bayesian Nash Equilibrium

A Bayesian Nash equilibrium (BNE) can be used for solving the chemical plant protection game if multiple types of attacker are involved in the game and when the attackers move, they do not have any information about the defender's defence plan. The BNE can capture the defender's uncertainties on the attacker's types, while the defender's uncertainties on the attacker's parameters/payoffs will be modelled in Chap. 4.

In the Bayesian CPP game, the defender's type is deterministic, and the attacker's type can vary. However, every type of attacker knows their own type and the defender knows a priori the probabilities of each type of attacker being involved.

To this end, the ex-interim Bayesian Nash Equilibrium should be employed. For more discussion on ex-ante, ex-interim, and ex-post Bayesian Nash equilibrium, interested readers are referred to Shoham and Leyton-Brown [15] and Ceppi et al. [16]

The ex-interim Bayesian Nash equilibrium $(\dot{y}, \dot{x}_1, \dot{x}_2, \ldots, \dot{x}_{|TL|})$ for the CPP game can be defined as shown in Formulas (3.23) and (3.24).

$$\dot{y} = \text{argmax}_{y \in Y} \left(\sum_{t \in TL} p^t \cdot \dot{x}_t^T \cdot U_d^t \right) \cdot y \tag{3.23}$$

$$\dot{x}_t = \text{argmax}_{x \in X_t} x_t^T \cdot U_a^t \cdot \dot{y}, t \in TL \tag{3.24}$$

In which U_a^t and U_d^t are the attacker and the defender's payoff matrix respectively, in case of an attacker type $t \in TL$; X_t is the mixed strategy space for attacker t; p^t is the prior probability of attacker t, as shown in Formulas (3.2).

The most well-known approach for solving a Bayesian game is by using the Harsanyi transformation, which transfers an incomplete information game (i.e., in the CPP game, with the multiple types of attackers) to a complete but imperfect information game. However, this approach computes the ex-ante Bayesian Nash equilibrium. For the ex-interim Bayesian Nash equilibrium, new algorithms are needed. Ceppi et al. [16] developed three algorithms for computing the interim BNE for two-player strategic form games, namely the B-PNS (based on support enumeration), the B-LC (based on linear complementarity formulation), and the B-SGC (based on mixed integer linear programming). It is worth noting that all these three algorithms have a high computational complexity, thus before implementing them, the dominance checking should be carried out on the game first.

3.3.4　Bayesian Stackelberg Equilibrium

A Bayesian Stackelberg equilibrium (BSE) can be employed to solve the chemical plant protection game, if multiple types of attackers are considered, and the attackers know the defender's defence plan when they attack.

The BSE $(\tilde{y}, \tilde{s}_a^1, \tilde{s}_a^2, \ldots, \tilde{s}_a^{|TL|})$ for the CPP game can be defined as shown in Formulas (3.25) and (3.26).

$$\tilde{y} = \text{argmax}_{y \in Y} \sum_{t \in TL} p^t \cdot U_d^t(\tilde{s}_a^t, :) \cdot y \tag{3.25}$$

$$\tilde{s}_a^t = \text{argmax}_{s_a^t \in S_a^t} U_a^t(s_a^t, :) \cdot y, \quad t \in TL \tag{3.26}$$

A straightforward approach for computing the BSE for the CPP game is to solve the linear programming problem for each combination of the attacker's best responses. This is quite similar to the MultiLP algorithm, and in total $\prod_{t \in TL} m^t$ linear programming problems need to be solved. Paruchuri et al. [17] proposed a

Table 3.2 The MovLib algorithm

Input:
A sequential CPP game with multiple types of attackers $\left(U_a^t, U_d^t, p^t\right), t \in TL$; an BSE of the game $\left(\tilde{y}, \tilde{s}_a^1, \tilde{s}_a^2, \ldots, \tilde{s}_a^{|TL|}\right)$, outputted from the DOBSS algorithm; a small constant positive value ε.
Output:
A Modified BSE $\left(\overline{\overline{y}}, \overline{\overline{s}}_a^1, \overline{\overline{s}}_a^2, \ldots, \overline{\overline{s}}_a^{|TL|}\right)$, or fail.
Solve the Linear Programming (LP):
$\begin{cases} \qquad Payoff = \max_{y \in Y} \sum_{t \in TL} p^t \cdot U_d^t\left(\tilde{s}_a^t, :\right) \cdot y \\ s.t. \ U_a^t\left(\tilde{s}_a^t, :\right) \cdot y \geq U_a^t\left(s_a^t, :\right) \cdot y + \varepsilon, \quad \forall s_a^t \in S_a^t - \left\{\tilde{s}_a^t\right\}, t \in TL \end{cases}$
If the LP feasible
$\qquad \overline{\overline{y}} \leftarrow$ the optimal point y, and $\overline{\overline{s}}_a^t \leftarrow \tilde{s}_a^t$;
$\qquad$ Return the Modified BSE $\left(\overline{\overline{y}}, \overline{\overline{s}}_a^1, \overline{\overline{s}}_a^2, \ldots, \overline{\overline{s}}_a^{|TL|}\right)$.
If the LP infeasible
$\qquad$ Return failure.

mixed integer linear programming (MILP) based algorithm for speeding up the computing of BSE for large scale Bayesian Stackelberg games, namely, the DOBSS algorithm. It is worth noting that the BSE calculated by the DOBSS algorithm is also a Strong Stackelberg Equilibrium, that is to say, the "breaking-tie" assumption (see Sect. 2.2.3) is applied.

Table 3.2 provides an algorithm named as 'MovLib' to slightly modify the BSE calculated by the DOBSS algorithm. The idea of the MovLib algorithm is that: based on the BSE, the defender moves a little bit from her BSE strategy, which is absolutely optimal for her but the "breaking-tie" assumption is required, to a strategy that is a bit less optimal to her but the "breaking-tie" assumption is no longer required.

Inputs for the MovLib are: payoff matrices for a sequential move CPP game with multiple types of attackers, i.e., U_a^t and U_d^t; the prior probability of the occurence of each type of attacker, i.e., p^t; a Bayesian Stackelberg Equilibrium of the CPP game, i.e., $\left(\tilde{y}, \tilde{s}_a^1, \tilde{s}_a^2, \ldots, \tilde{s}_a^{|TL|}\right)$, which can be calculated by the DOBSS algorithm; a small constant positive value ε, which can be, for instance, 0.1. The output of the MovLib algorithm is a Modified BSE $\left(\overline{\overline{y}}, \overline{\overline{s}}_a^1, \overline{\overline{s}}_a^2, \ldots, \overline{\overline{s}}_a^{|TL|}\right)$. The main body of the algorithm is the Linear Programming (LP) problem. The cost function of the LP denotes that the defender is optimizing her payoff, knowing that the attacker t would play his strategy $\tilde{s}_a^t$. The constraint of the LP makes sure that the attacker t's payoff by playing strategy $\tilde{s}_a^t$ is at least ε more than his payoff by playing any other strategies. The constraint makes sure that $\tilde{s}_a^t$ is the unique best response strategy for the attacker and therefore there is no "breaking-tie" problem anymore.

The MovLib algorithm can fail, if there is a $\tilde{s}_a^t$ that for any $y \in Y$, the constraint of the LP in the MovLib algorithm can not hold. For instance, if attacker t has a strategy s_a^t that has exactly the same payoff (no matter what the defender's strategy is) as strategy $\tilde{s}_a^t$, then the constraint in the algorithm would not hold, no matter how small the ε is.

3.4 CPP Game from an Industrial Practice Point of View

Developing, solving and using the chemical plant protection game needs massive quantitative inputs and the output of the CPP game is also quantitative. In this section, discussions on how to obtain these inputs via conventional security risk assessment methodologies and on how to translate the game theoretic outputs to be usable for these conventional approaches, are given.

The American Petroleum Institute (API) recommends to employ Security Risk Assessment (henceforth, the "API SRA") methodologies in the petroleum and petrochemical industries. The API SRA indeed provides a systematic and practically implementable framework for security risk analysis in the process industries. Since first published in 2004, it was re-edited in 2013, and it has been extensively used in industrial practice. In this section, we choose the API SRA as the baseline methodology, and show how to obtain inputs for the CPP game from the API SRA and to translate the game theoretic results back to the API SRA terminologies. For information about the API SRA methodology, please see Sect. 1.3.3 in this book, or see [5].

3.4.1 Input Analysis

The following inputs are needed for calculating the payoff matrices for the CPP game: (i) the prior probabilities of each type of attackers (i.e., p^t); (ii) the probabilities that the intruder can pass an entrance (i.e., p_{rj}^P) or a zone (i.e., p_i^z), under certain defence and certain attack scenario; (iii) conditional probabilities that an attack will be implemented if the attacker has successfully reached the target (i.e., P_y); (iv) estimated consequences of a successful attack (i.e., L); and (v) the defence and the attack costs, say, the C_d and C_a, respectively.

The prior probabilities of each type of attacker can be calculated according to the threat level linked to each attacker, as was discussed in Sect. 3.2.1. In the API SRA methodology, the SRA team first decides what kind of threat the plant is faced with, and obtains the threat list TL. The team further estimates a threat level for each type of threat, according to the criteria shown in Table 3.3 (adopted from the API document [5]), and obtains the threat score list TS. Based on TL and TS, the prior probabilities of each attacker can be calculated based on Formula (3.2) (see Sect. 3.2.1).

The probability that an intruder can successfully reach the target depends on the defender's defence scenario and on the attacker's intrusion path, and thus, it is a function of both the defender and the attacker's actions. For instance, the probability would be quite low if the defender deploys an x-ray scanner at the main entrance, and the attacker chooses to intrude with an explosive device; while the probability could be higher if the attacker wants to implement an attack by switching off a key safety valve in the plant since the x-ray scanner does not work on this attack scenario.

Table 3.3 Threat ranking criteria

API SRA Methodology	
Threat level	Description[a]
1—Very low	Indicates little or no credible evidence of capability or intent and no history of actual or planned threats against the asset or similar assets (e.g. "no expected attack in the life of the facility's operation").
2—Low	Indicates that there is a low threat against the asset or similar assets and that few known adversaries would pose a threat to the asset (e.g. "1 event or more is possible in the life of the facility's operation").
3—Medium	Indicates that there is a possible threat to the asset or similar assets based on the threat's desire to compromise similar assets, but no specific threat exists for the facility or asset (e.g. "1 event or more in 10 years of the facility's operation is possible").
4—High	Indicates that a credible threat exists against the asset or similar assets based on knowledge of the threat's capability and intent to attack the asset or similar assets, and some indication exists of the threat specific to the company, facility, or asset (e.g. "1 event or more in 5 years of the facility's operation is possible").
5—Very high	Indicates that a credible threat exists against the asset or similar assets; that the threat demonstrates the capability and intent to launch an attack; that the subject asset or similar assets are targeted or attacked on a frequently recurring basis; and that the frequency of an attack over the life of the asset is very high (e.g. "1 event/ event per year is possible").

[a]User defined values should be applied

Table 3.4 (adopted from the API document [5]) shows how the API SRA methodology quantifies these probabilities.

The conditional probability of a successful attack and the estimated consequences depend only on the attacker's strategy, that is, which facility he wants to attack, and with what scenario. For instance, to cause losses from an explosion on an oil storage tank with an explosive device can be easier than to cause them by switching off a key safety valve, since the latter scenario needs more professional knowledge. Moreover, different attack scenarios will of course result in different consequences.

In the API SRA methodology, there is no separated assessment of the intrusion probabilities and the conditional success probabilities. Table 3.4 may also be used for obtaining the conditional success probability. For the consequences, the API SRA uses a ranking method to measure the consequences of an event. However, the scores that an event will receive are based on quantitative descriptions, as shown in Table 3.5 (adopted from the API document [5]). The idea is to use the quantitative data from the left-hand column directly, instead of using the scores in the right-hand column. There are 5 different aspects of the quantitative data, namely, the casualties, the environment impacts, direct economic loss, business interruption, and reputation impacts. A set of coefficients are needed to utilize them into monetary numbers. For instance, to transfer a casualty as 5.8 million euro in the Netherlands [18, 19].

The defence and the attack costs depend on the defender's defence plan and on the attacker's attack scenario, respectively. Generally speaking, higher security alert levels at each entrance and in each zone will secure the plant better, however, the

Table 3.4 Vulnerability scores and corresponding quantitative data

API SRA Methodology			
VL[a]	D	CPS	Description
1	Very low	[0.0, 0.2]	Indicates that multiple layers of effective security measures to deter, detect, delay, respond to, and recover from the threat exist, and the chance that the adversary would be readily able to succeed at the act is very low.
2	Low	(0.2, 0.4]	Indicates that there are effective security measures in place to deter, detect, delay, respond, and recover; however, at least one weakness exists that a threat would be able to exploit with some effort to evade or defeat the countermeasures.
3	Medium	(0.4, 0.6]	Indicates that although there are some effective security measures in place to deter, detect, delay, respond, and recover, but there is not a complete and effective application of these security strategies and so the asset or the existing countermeasures could still be compromised.
4	High	(0.6, 0.8]	Indicates there are some security measures to deter, detect, delay, respond, and recover, but there is not a complete or effective application of these security strategies and so the adversary could succeed at the act relatively easily.
5	Very high	(0.8, 1.0]	Indicates that there are very ineffective security measures currently in place to deter, detect, delay, respond, and recover, and so the adversary would easily be able to succeed.

[a]VL vulnerability level, D descriptor, CPS conditional probability of success

defence cost will also be higher. A well trained attacker will always increase the success probability of an attack scenario as well as increase the consequence of the attack. In the API SRA, at the mitigation step (step 5.1 in the API SRA document), it is clearly mentioned that the costs of mitigation options should be considered. Hence, it is possible to obtain the C_d for the defender. For the attack costs, the API SRA does not point out how to estimate it. However, the SRA team could make an estimation of the attacker's cost based on an attack scenario. For instance, different types of bombs, different bombers (well-trained or not), different vehicles related to different attack scenarios, and a different scenario related to a different attack cost, can be conceptualized and mapped.

Though the API SRA framework provides some data for the CPP game, some more work is needed. The API SRA focuses on estimating threats/probabilities/consequences from the defender's view point. However, it is not necessary that the attacker and the defender have the same preference, which was already shown in the payoff definitions (see, Formulas (3.7) and (3.8)) of the CPP game. For example, an important facility might not be an attractive target to an attacker, since the attacker has his own preference. The attacker would plan his attack according to his own preference and using his estimation of the defender's strategy. Moreover, the API SRA framework does not explicitly evaluate if the attacker would know her defence plan or not. In a game theoretic terminology, if the attacker knows the defender's strategy, the game is said to being played sequentially, and a (Bayesian) Stackelberg equilibrium should therefore be used. Otherwise, if the attacker would not know, the

Table 3.5 Consequence ranking and the corresponding quantitative data

API SRA Methodology	
Description	Ranking
(a) Possibility of minor injury on-site; no fatalities or injuries anticipated off site.	1
(b) No environmental impacts.	
(c) Up to $X loss in property damage.	
(d) Very short-term (up to X weeks) business interruption/expense.	
(e) Very low or no impact or loss of reputation or business viability; mentioned in local press.	
(a) On-site injuries that are not widespread but only in the vicinity of the incident location; no fatalities or injuries anticipated off site.	2
(b) Minor environmental impacts to immediate incident site area only, less than X year(s) to recover.	
(c) $X to $X loss in property damage.	
(d) Short-term (>X week to Y months) business interruption/expense.	
(e) Low loss of reputation or business viability; query by regulatory agency; significant local press coverage.	
(a) Possibility of widespread on-site serious injuries; no fatalities or injuries anticipated off site.	3
(b) Environmental impact on-site and/or minor off-site impact, Y year(s) to recover.	
(c) Over $X to $X loss in property damage.	
(d) Medium-term (Y to Z months) business interruption/expense.	
(e) Medium loss of reputation or business viability; attention of regulatory agencies; national press coverage.	
(a) Possibility of X to Y on-site fatalities; possibility of off-site injuries.	4
(b) Very large environmental impact on-site and/or large off-site impact, between Y and Z years to recover.	
(c) Over $X to $X loss in property damage.	
(d) Long-term (X to Y years) business interruption/expense.	
(e) High loss of reputation or business viability; prosecution by regulator; extensive national press coverage.	
(a) Possibility of any off-site fatalities from large-scale toxic or flammable release; possibility of multiple on-site fatalities.	5
(b) Major environmental impact on-site and/or off site (e.g. large-scale toxic contamination of public waterway), more than XX years/poor chance of recovery.	
(c) Over $X loss in property damage.	
(d) Very long-term (>X years) business interruption/expense; large-scale disruption to the national economy, public or private operations; loss of critical data.	
(e) Very high loss of reputation or business viability; international press coverage.	

game is said to being played simultaneously, and a (Bayesian) Nash equilibrium therefore should be employed.

Instead of being obtained via security experts, the vulnerability related inputs (e.g., p_i^z) can also be calculated by a function named the "Contest Success Function (CSF) [4], for theoretical study purpose. The CSF investigates the success probability of each player in a multiple players contest. The success probability is determined both by the player and his opponents' contest effort and can be further formulated as:

$$p^i(e) = {}^{\alpha_i \cdot e_i^r}\!\Big/\!\!\sum_{j \in N} {}^{\alpha_j \cdot e_j^r} \qquad (3.27)$$

In which e denotes the players' effort; $r > 0$ and $\alpha > 0$ are constant real numbers which should be determined by the contest circumstance. For more information on the CSF, readers are referred to Clark and Rijs [20], Skaperdas [4], and Guan and Zhuang [21].

3.4.2 Output Analysis

The Chemical Plant Protection game outputs an equilibrium strategy pair (s) $\left(\bar{s}_d, \bar{s}_a^t\right)$ (might also be a mixed strategy pair) and a corresponding equilibrium payoff (s) $\left(\bar{u}_d, \bar{u}_a^t\right)$. In this section we show how to translate these outputs to the API SRA terminologies.

The defender's strategy is modelled as setting security alert levels at each entrance and zone. In industrial practice, different security alert levels represent different combinations of security countermeasures (see Table 3.1). Therefore, the equilibrium strategy $\bar{s}_d$ can be mapped to the proposed countermeasures list CML in the API SRA methodology. If the equilibrium strategy is a mixed strategy, which denotes the probabilities of setting different security alert levels, then it can be mapped to the prioritization procedure of the countermeasures list. Furthermore, the countermeasures with a higher mixed strategy probability should be assigned a higher priority.

The attacker's equilibrium strategy $\bar{s}_a^t$ clearly indicates which target the attacker will attack and what attack scenario will be employed. It is counter-intuitive that when knowing this, the defender wouldn't enhance the protection of the attacker's target. Nonetheless, this is a result of the intelligent interactions between the defender and the attacker: if the defender protects the equilibrium target better, then the attacker would also deviate from his current target to a new target.

The defender's equilibrium payoff $\bar{u}_d$ reflects the mitigated security risk, and in the API SRA methodology, it is denoted as R^2. The attacker's equilibrium payoff $\bar{u}_a^t$ reflects the attacker's attack motivation, and in the API SRA methodology, it is named the "degree of interest".

It is worth noting that the attacker's equilibrium strategy can also be a mixed strategy. Denote the likelihood that target tgt would be attacked as: $lk_{tgt} = \sum_{i \in M}\,_{(tgt)} x_i$, in which $M(tgt) \subset M$ denotes all the attacker pure strategies that take tgt as the attack target. In the API SRA methodology, lk_{tgt} is represented as "attractiveness", as shown in Table 3.6 (adopted from the API document [5]).

Table 3.6 Attractiveness ranking level

API SRA Methodology			
RL[a]	D	CPA	Threat ranking
1	Very low	[0.0, 0.2]	Threat would have little to no level of interest in the asset.
2	Low	(0.2, 0.4]	Threat would have some degree of interest in the asset, but it is not likely to be of interest compared to other assets.
3	Medium	(0.4, 0.6]	Threat would have a moderate degree of interest in the asset relative to other assets.
4	High	(0.6, 0.8]	Threat would have a high degree of interest in the asset relative to other assets.
5	Very high	(0.8, 1.0]	Threat would have a very high degree of interest in the asset, and it is a preferred choice relative to other assets.

[a]RL ranking level, D descriptor, CPA conditional probability of the act

3.5 Conclusion

In this chapter, the most basic form of the chemical plant protection (CPP) game is proposed. The CPP game as put forward and explained in this chapter assumes complete information and rational players. Several typical game solutions are defined for the game, and the user of the game may decide which solution to use according to his/her belief of the threat. If only one threat is possible, and the threat is not able to know the defender's defence plan, then a NE should be used; if only one threat is possible, and the threat knows the defender's plan, then a SE should be used; if multiple threats exist, a Bayesian simultaneous CPP game or a Bayesian Stackelberg CPP game should be used, in the case that the attacker does not know, and knows, the defender's plant, respectively.

Inputs and outputs of the game are also discussed. The CPP game needs a lot of quantitative input data, and conventional security risk analysis approaches (e.g., the API SRA) are able to provide this required information. The output of the game can also be mapped to concepts used in conventional security risk analysis methodologies.

Drawbacks of the basic CPP game are obvious. Firstly, only the uncertainty of different types of attackers are modelled, while the uncertainties of the attacker's parameters/information and the uncertainties of the attacker's rationalities are not considered. Chapters 4 and 5 will address these two types of uncertainties respectively. Secondly, the CPP game works only for intrusion attacks. For remote attacks (i.e., through a network attack on the plant's control system) the CPP game cannot be employed since such attacks have totally different characteristics compared to intrusions. Hence, for cyber security new models are needed. Also, the exit procedure of an intrusion attack is not considered, which in case of a thief threat can be quite important. Thirdly, industrial security practice is more complicated than our theoretical description. For example, the mixed strategy for the defender is explained as the defender changing her security alert level day to day. However, in practice, the defender may not be able to change the SAL because of some location-fixed equipment (e.g., camera system).

References

1. Zhang L, Reniers G. A game-theoretical model to improve process plant protection from terrorist attacks. Risk Anal. 2016;36(12):2285–97.
2. Zhang L, Reniers G. Applying a Bayesian Stackelberg game for securing a chemical plant. J Loss Prev Process Ind. 2018;51:72–83.
3. Zhang L, Reniers G, Chen B, Qiu X. Integrating the API SRA methodology and game theory for improving chemical plant protection. J Loss Prev Process Ind. 2018;51(Suppl C):8–16.
4. Skaperdas S. Contest success functions. Econ Theory. 1996;7(2):283–90.
5. API. Security risk assessment methodology for the petroleum and petrochemical industries. In: 780 ARP, editor. 2013.
6. Brown G, Carlyle M, Salmerón J, Wood K. Defending critical infrastructure. Interfaces. 2006;36(6):530–44.
7. Baybutt P. Issues for security risk assessment in the process industries. J Loss Prev Process Ind. 2017;49(Part B):509–18.
8. Talarico L, Reniers G, Sörensen K, Springael J. MISTRAL: a game-theoretical model to allocate security measures in a multi-modal chemical transportation network with adaptive adversaries. Reliab Eng Syst Saf. 2015;138:105–14.
9. Moulin H, Vial J-P. Strategically zero-sum games: the class of games whose completely mixed equilibria cannot be improved upon. Int J Game Theory. 1978;7(3–4):201–21.
10. Lemke CE, Howson J, Joseph T. Equilibrium points of bimatrix games. J Soc Ind Appl Math. 1964;12(2):413–23.
11. Rao NS, Poole SW, Ma CY, He F, Zhuang J, Yau DK. Defense of cyber infrastructures against cyber-physical attacks using game-theoretic models. Risk Anal. 2015;36:694–710.
12. Zhuang J, Bier VM. Balancing terrorism and natural disasters-defensive strategy with endogenous attacker effort. Oper Res. 2007;55(5):976–91.
13. Gibbons R. A primer in game theory. New York: Harvester Wheatsheaf; 1992.
14. Conitzer V, Sandholm T, editors. Computing the optimal strategy to commit to. In: Proceedings of the 7th ACM conference on electronic commerce; 2006: ACM.
15. Shoham Y, Leyton-Brown K. Multiagent systems: algorithmic, game-theoretic, and logical foundations. Cambridge: Cambridge University Press; 2008.
16. Ceppi S, Gatti N, Basilico N, editors. Computing Bayes-Nash equilibria through support enumeration methods in Bayesian two-player strategic-form games. In: Proceedings of the 2009 IEEE/WIC/ACM international joint conference on web intelligence and intelligent agent technology-Volume 02; 2009: IEEE Computer Syociety.
17. Paruchuri P, Pearce JP, Marecki J, Tambe M, Ordonez F, Kraus S, editors. Playing games for security: an efficient exact algorithm for solving Bayesian Stackelberg games. In: Proceedings of the 7th international joint conference on autonomous agents and multiagent systems-Volume 2; 2008: international foundation for autonomous agents and multiagent systems.
18. IENM. Letter of 26 April 2013 to the parliament. In: 2013/19920 MvIIB, editor. 2013.
19. Reniers G, Van Erp H. Operational safety economics: a practical approach focused on the chemical and process industries. Chichester: Wiley; 2016.
20. Clark DJ, Riis C. Contest success functions: an extension. Econ Theory. 1998;11(1):201–4.
21. Guan P, Zhuang J. Modeling resources allocation in attacker-defender games with "Warm Up" CSF. Risk Anal. 2016;36(4):776–91.

Chapter 4
Single Plant Protection: Playing the Chemical Plant Protection Game with Distribution-Free Uncertainties

In this chapter, the Chemical Plant Protection game is extended to deal with input parameters with distribution-free uncertainties [1] The so-called interval CPP game is defined. Two algorithms, namely, the interval bi-matrix game solver (IBGS) and the interval CPP game solver (ICGS), are proposed.

4.1 Motivation

The defender plans her defence according to her guess about the attacker's behaviour, while the attacker also plans and implements his attack based on his information on the defender's defence. The CPP game models these intelligent interactions between the company defender and the potential attackers. Uncertainties on multiple types of attackers may also be modelled by using the Bayesian Nash/Stackelberg equilibrium. Inputs of the game can be obtained by using some conventional security risk assessment methods such as the API SRA, and the outputs of the game can be translated to the conventional security risk analysis terminologies.

However, in industrial practice, as well as in most conventional security risk assessment methods, it is quite difficult to obtain exact numbers (point values) for most parameters that the CPP game needs. For instance, in Table 3.4 (see Sect. 3.4.1), the SRA team may only provide an interval of the consequences, such as a property loss from 1 k euro to 2 k euro. Furthermore, the team would not be able to know how the consequences are distributed on the interval, and whether the consequences are uniformly distributed between [1000, 2000] euro, or whether they follow a triangle shape between this interval, among other possibilities. The

---

This chapter is based on the published paper of Zhang et al. [1].

© Springer International Publishing AG, part of Springer Nature 2018
L. Zhang, G. Reniers, *Game Theory for Managing Security in Chemical Industrial Areas*, Advanced Sciences and Technologies for Security Applications, https://doi.org/10.1007/978-3-319-92618-6_4

estimation on vulnerabilities (i.e., intrusion probabilities and conditional probabilities of a successful attack) is even more difficult than the estimation of consequences.

There are three approaches to deal with interval inputs with distribution-free uncertainties. The first approach is to use a figure representing the interval, such as using the median or an average number. By using this approach, we obtain the inputs for the CPP game directly, and then solve the game. The result of this approach is not robust. The CPP game is finally solved by (Mixed Integer) Linear Programming (LP), no matter which equilibrium concept is employed. It is well-known that the optimal value of a Linear Programming problem is always situated on the boundary of the feasible area [2]. Thus a small error on the input parameters would make the LP result very bad [3], and in the CPP game, the "real exact" number (which we do not know) would always have an (at least small) error to the representative number that we use. The second approach is to add assumptions on the parameters' distribution on the interval, and then solve the game with continuous uncertainties. For instance, assume that the consequence is uniformly distributed between [1000, 2000]. With these assumptions, several algorithms can be employed to solve the game, see for instance, a comprehensive investigation by Kiekinveld et al. [4]. An obvious drawback of this approach is the assumption on the distribution. Different distribution assumptions may have different results and in practice it is difficult to decide which distributions to use. This approach is also computationally quite time-consuming. The third approach is by employing robust optimization techniques, and works directly on the distribution-free uncertainties, see for instance, Kiekintveld et al.[5] and Nikoofal and Zhuang [6]. In this approach, no extra assumption is needed. The defender knows the intervals that the attacker's parameters will be situated in, and she plays the game conservatively thinking that the attacker's parameters are located at the worst point (but still within the interval) for her. In this chapter, the third approach is employed and explained.

As discussed in Sect. 3.3.2, a sequential CPP game may bring the defender a "First-Mover advantage", and it does not have the Equilibrium choice problem. Moreover, a sequential CPP game also reflects industrial reality better. Therefore, in this chapter and in Chap. 5, we limit our research to the sequential CPP game.

4.2   Interval CPP Game Definition

Recalling the payoff Formula (3.8) (see Sect. 3.2.3) and the intrusion probability calculation Formula (3.1) (see Sect. 3.1), the attacker's payoff can be re-written as Formula (4.1).

$$u_a(s_a, s_d) = \prod_{i=0}^{I} \tilde{P}_i^z \cdot \prod_{j=1}^{I} \tilde{P}_j^p \cdot \widetilde{P_y}(s_a) \cdot \tilde{L}_y(s_a) - C_a(s_a) \qquad (4.1)$$

For each attacker parameter (i.e., the $\tilde{P}, \widetilde{P_y}, \tilde{L}_y, C_a$), for the sake of convenience, denoting it as σ, assume that the defender does not know the exact value of σ and she knows that $\sigma \in [\sigma^{min}, \sigma^{max}]$. Since $\tilde{P}_i^z, \tilde{P}_j^p, \widetilde{P_y}, \tilde{L}_y \geq 0$, we can easily derive that:

$$u_a^{min} = \prod\nolimits_{i=0}^{I} \tilde{P}_i^z min \cdot \prod\nolimits_{j=1}^{I} \tilde{P}_j^p min \cdot \tilde{P}_y^{\,min} \cdot \tilde{L}_y^{\,min} - C_a^{\,max} \qquad (4.2)$$

$$u_a^{max} = \prod\nolimits_{i=0}^{I} \tilde{P}_i^z max \cdot \prod\nolimits_{j=1}^{I} \tilde{P}_j^p max \cdot \tilde{P}_y^{\,max} \cdot \tilde{L}_y^{\,max} - C_a^{\,min} \qquad (4.3)$$

In this research, the defender is assumed to know the exact numbers of her own parameters. Therefore, in the Interval CPP game, the defender's payoff matrix is the same as in the CPP game, while the attacker's payoff matrix consists of an upper bound matrix and a lower bound matrix, as defined in Formulas (4.2) and (4.3). We denote the Interval CPP game as $ICG = \{U_d, U_a^{min}, U_a^{max}\}$.

4.3   Interval Bi-Matrix Game Solver (IBGS)

In the Interval CPP game, if the defender commits to a mixed strategy $y \in Y$, she would not be able to work out the attacker's best response, due to the existence of uncertainties. Contrary to Formula (3.20) (see Sect. 3.3.2), the defender in an interval game only knows an interval of the attacker's payoffs related to responding to a pure strategy, denoted as $u_a^i \in [u_a^{i-min}, u_a^{i-max}]$, and we have $u_a^{i-min} = U_a^{min}(i, :) \cdot y$ and $u_a^{i-max} = U_a^{max}(i, :) \cdot y$.

Knowing the range of the attacker's payoffs, the defender can work out the attacker's maximal lower bound of payoffs among all the attacker's pure strategies, i.e., $R = \max_{i \in M} u_a^{i-min}$, in which $M = \{1, 2, \dots, m\}$ and $i \in M$ means for each attacker pure strategies (i.e., $s_a \in S_a$). The defender beliefs that a rational attacker would not play a strategy whose upper bound payoff is less than R. Formula (4.4) illustrates the reason for this judgement, in which ml is the strategy who has the maximal lower bound payoff, k is the strategy whose upper bound payoff is less than R. The formula shows that the attacker would always have a higher payoff by responding with strategy ml instead of by responding with strategy k, to the defender's strategy y.

$$U_a(k, :) \cdot y \leq u_a^{k-max} < R = u_a^{ml-min} \leq U_a(ml, :) \cdot y \qquad (4.4)$$

Based on the above analysis, an algorithm for solving the Chemical Plant Protection game with distribution-free uncertainties is proposed, as shown in Formula (4.5).

$$\max_{q,h,\gamma,y,R} \sum_{t \in TL} p^t \gamma^t$$

$$s.t. \begin{cases} c1. \ 0 \le R^t - \underline{U}_a^t(i,:) \cdot y \le \left(1 - h_i^t\right) \cdot \Gamma, \ \forall i \in M^t \\ c2. \ \left(q_i^t - 1\right) \cdot \Gamma \le \bar{U}_a^t(i,:) \cdot y - R^t \le q_i^t \cdot \Gamma, \ \forall i \in M^t \\ c3. \ \Gamma \cdot \left(1 - q_i^t\right) + U_d^t(i,:) \cdot y \ge \gamma^t, \ \forall i \in M^t \\ c4. \ q_i^t \ge h_i^t, \ \forall i \in M^t \\ c5. \ q_i^t, h_i^t \in \{0,1\} \\ c6. \ \sum h^t = 1 \\ c7. \ \sum y = 1, y_i \in [0,1] \\ c8. \ R^t, \gamma^t \in R \end{cases} \quad (4.5)$$

In the algorithm, t, TL, p^t denote a threat, the threat list, and the prior probability of threat, respectively, as already defined in previous chapters; R^t is the maximal value of the lower bound payoffs for threat t; $\underline{U}_a^t$ and $\bar{U}_a^t$ are the lower bound and upper bound payoff matrices for threat t respectively; $M^t = \{1, 2, \ldots, m^t\}$ is the pure strategy index set; h_i^t and q_i^t are binary variables; U_d^t denotes the defender's payoff matrix in case of a threat t; Γ is a constant large real number.

To understand the algorithm, notice that in constraint $c1$, if $h_i^t = 1$, we obtain $R^t = \underline{U}_a^t(i,:)?y$, otherwise if $h_i^t = 0$, we obtain $R^t \ge \underline{U}_a^t(i,:) \cdot y$. Thus the binary variable h_i^t represents that whether the ith pure strategy of threat t has the maximal lower bound payoff, and if yes, $h_i^t = 1$, if no, $h_i^t = 0$. In constraint $c2$, if $q_i^t = 1$, we get $R^t \le \bar{U}_a^t(i,:) \cdot y$, otherwise if $q_i^t = 0$, we have $R^t \ge \bar{U}_a^t(i,:) \cdot y$. To this end, the binary variable q_i^t denotes that whether the ith pure strategy of threat t has a higher upper bound payoff than R^t, and if yes, $q_i^t = 1$, if no, $q_i^t = 0$. Constraint $c3$ is activated only when $q_i^t = 1$, which means that the ith strategy is a possible choice for the attacker t. $c3$ together with the cost function also indicate that among all the possible attacker strategies, the defender conservatively aims to optimize the worst case γ^t. Constraint $c4$ means that the strategy who has the highest lower bound payoff must be a possible strategy for the attacker. To understand $c4$, one must notice that (i) in $c2$, if $R^t = \bar{U}_a^t(i,:) \cdot y$, which means that the attacker's upper bound payoff by playing strategy i exactly equals R^t, then q_i^t can be either 0 or 1; (ii) if q_i^t can be either 0 or 1, then in some cases, constraint $c3$ in continuation with the cost function will lead to a result that $q_i^t = 0$; (iii) in some special situations, for instance, in a situation that the interval radius equals 0, which means that $\underline{U}_a^t = \bar{U}_a^t$, constraint $c4$ is needed to make sure that the strategy which will bring the attacker the highest lower bound payoff will be a possible best response strategy.

We may notice that Formula (4.5) does not use any special characteristics of the CPP game. Indeed, it is applicable to any bi-matrix games with distribution-free uncertainties. Therefore, we name it as Interval Bi-Matrix Game Solver (IBGS). It is also worth noting that this algorithm can be used for solving games with bounded rational attackers who are $\epsilon - optimal$ players, more details will be given in Sect. 5.2 of this book. The IBGS is mainly derived from the BRASS algorithm developed by Pita et al. [7].

Table 4.1 Hypothetical parameters for illustrating the parameter coupling problem	Strategy s_{a1}			Strategy s_{a2}		
	Para	Min	Max	Para	Min	Max
	$\tilde{P}_0^z$	0.8	0.9	$\tilde{P}_0^z$	0.8	0.9
	C_a	10	12	C_a	10	12
	$\tilde{P}_y$	0.9	0.92	$\tilde{P}_y$	0.8	0.84
	$\tilde{L}_y$	100	102	$\tilde{L}_y$	130	140

4.4 Parameter Coupling

The IBGS, being general enough, loses some model properties of the CPP game. Recalling Formulas (4.2) and (4.3), the uncertainties on the parameters result in uncertainties on the payoffs. However, different attacker strategies may share some parameters, and the shared parameters cannot reach their maximal values in one strategy while reaching their minimal values in another.

An illustrative example can be useful to make the above statement clear. Assume that attacker strategies s_{a1} and s_{a2} aim to attack different targets with the same attack scenario, and both targets are situated in zone level 0 (i.e., outside of the plant, for simplicity reasons). In this case, the attacker's payoff can be simplified as $u_a(s_a, s_d) = \tilde{P}_0^z(s_a, s_d) \cdot \widetilde{P}_y(s_a) \cdot \tilde{L}_y(s_a) - C_a(s_a)$. Further assume that the defender plays a pure strategy s_d resulting in the parameters as shown in Table 4.1. The only difference between these two strategies is the target, thus we can see that in Table 4.1, $\tilde{P}_0^z(s_{a1}, s_d)$ and $\tilde{P}_0^z(s_{a2}, s_d)$, as well as $C_a(s_{a1})$ and $C_a(s_{a2})$, have the same range. The target related parameters $\widetilde{P}_y$ and $\tilde{L}_y$, conversely, have different ranges.

Based on Formulas (4.2) and (4.3), we have:

$$u_a^{min}(s_{a1}, s_d) = 0.8 \cdot 100 \cdot 0.9 - 12 = 60 \tag{4.6}$$
$$u_a^{max}(s_{a1}, s_d) = 0.9 \cdot 102 \cdot 0.92 - 10 = 74.456 \tag{4.7}$$
$$u_a^{min}(s_{a2}, s_d) = 0.8 \cdot 130 \cdot 0.8 - 12 = 71.2 \tag{4.8}$$
$$u_a^{max}(s_{a2}, s_d) = 0.9 \cdot 140 \cdot 0.84 - 10 = 95.84 \tag{4.9}$$

According to algorithm IBGS, we have $u_a^{min}(s_{a2}, s_d) = R \le u_a^{max}(s_{a1}, s_d)$. Therefore, the defender may not be able to know whether the attacker would play strategy s_{a1} or not. However, strategies s_{a1} and s_{a2} have the same parameters $\tilde{P}_0^z$ and C_a. We substitute other parameters (i.e., $\widetilde{P}_y$ and $\tilde{L}_y$) into Formulas (4.2) and (4.3), and remain the shared parameters (i.e., $\tilde{P}_0^z$ and C_a), we obtain:

$$u_a^{min}(s_{a1}, s_d) = 90 \cdot \tilde{P}_0^z - C_a \tag{4.10}$$
$$u_a^{max}(s_{a1}, s_d) = 93.84 \cdot \tilde{P}_0^z - C_a \tag{4.11}$$
$$u_a^{min}(s_{a2}, s_d) = 104 \cdot \tilde{P}_0^z - C_a \tag{4.12}$$
$$u_a^{max}(s_{a2}, s_d) = 117.6 \cdot \tilde{P}_0^z - C_a \tag{4.13}$$

Although the defender only knows the intervals where the parameters $\tilde{P}_0^z$ and C_a are located in, she knows that $93.84 \cdot \tilde{P}_0^z - C_a = u_a^{max}(s_{a1}, s_d) \leq u_a^{min}(s_{a2}, s_d) = 104 \cdot \tilde{P}_0^z - C_a$. Based on this information, the defender can conclude that, for the attacker, strategy s_{a2} is always a better strategy than strategy s_{a1}. However, algorithm IBGS cannot support the defender to draw this conclusion. The reason is that, in the IBGS, when calculating the attacker's lower bound payoff by playing strategy s_{a2} (see Formula (4.8)), $\tilde{P}_0^z$ is substituted by 0.8 and C_a is substituted by 12 (see Table 4.1, illustrative figure), while when calculating the attacker's upper bound payoff by playing strategy s_{a1} (see Formula (4.7)), $\tilde{P}_0^z$ is substituted by 0.9 and C_a is substituted by 10 (see Table 4.1, illustrative figure). However, strategies s_{a1} and s_{a2} use the same attack scenario and their paths in zone 0 are the same. Therefore, the attack cost (i.e., C_a) of these two strategies, and the probabilities of being detected in zone 0 of these two strategies (i.e., $\tilde{P}_0^z$), should be the same, although the defender does not know the exact numbers of these parameters. Hence, when calculating the lower bound payoff, the lower bound values of the shared parameters (i.e., $\tilde{P}_0^z$ and C_a) are used, while when calculating the upper bound payoff, the upper bound values are used, and the fact is that the shared parameters should not reach their lower bound in one strategy and reach their upper bound in another strategy.

To formulate the parameters coupling problem as illustrated above, the so-called attacker payoff differences of two attacker pure strategies should be defined, as shown in Formula (4.14).

$$\Delta_{kl} = U_a(k,:) \cdot y - U_a(l,:) \cdot y, \forall k, l \in M \qquad (4.14)$$

Further define Tp_k as the set of sub-zones and entrances that attacker pure strategy k would pass, and define Tp_l analogously. Define $Tp_{k \cdot l} = Tp_k \cap Tp_l$, denoting zones or entrances that the two strategies both pass by. For instance, if the attacker would follow an intrusion path as p1 in Fig. 3.1 (see Sect. 3.1), then we have:

$$Tp_1 = \{zone0, truck\ entrance, zone\ 1\_1, entrance\ 3, zone\ 2\_2\},$$

and if the p2 in Fig. 3.1 would be followed (only for illustrating Tp purpose), we obtain:

$$Tp_2 = \{zone0, main\ entrance, zone\ 1\_1, entrance\ 1, zone\ 2\_1, entrance\ 2\};$$

and,

$$Tp_{1 \cdot 2} = \{zone\ 0, zone\ 1\_1\}.$$

Substitute payoff Formula (4.1) into Formula (4.14), resulting in:

$$\Delta_{kl} = \sum\nolimits_{j\in N} \left(\prod\nolimits_{i\in Tp_k} \tilde{p}_{ij} \cdot \widetilde{PL}_k - C_k\right) \cdot y_j$$

$$- \sum\nolimits_{j\in N} \left(\prod\nolimits_{i\in Tp_l} \tilde{p}_{ij} \cdot \widetilde{PL}_l - C_t\right) \cdot y_j \qquad (4.15)$$

In which $N = \{1, 2, \ldots, n\}$ denotes the defender's pure strategy index space; $\tilde{p}_{ij}$ denotes the success probability of passing an entrance i or a sub-zone i (which must be on the attacker's intrusion path) if the defender plays a pure strategy j; $\widetilde{PL}_k$ represents the conditional expected loss in case the attacker already arrived at the target, i.e., $\widetilde{PL}_k = \tilde{P}_k \cdot \tilde{L}_k$; C_k is the cost of attacker strategy k; y_j is the probability that the defender plays pure strategy j.

Formula (4.15) can be further organized as:

$$\Delta_{kl} = \sum\nolimits_{j\in N} \left[\left(\prod\nolimits_{i\in Tp_{k\cdot l}} \tilde{p}_{ij}\right) \cdot \left(\prod\nolimits_{i\in Tp_k - Tp_{k\cdot l}} \tilde{p}_{ij} \cdot \widetilde{PL}_k - \prod\nolimits_{i\in Tp_l - Tp_{k\cdot l}} \tilde{p}_{ij} \cdot \widetilde{PL}_l\right) \right] \cdot y_j$$
$$+ C_l - C_k$$

$$(4.16)$$

Which separates the shared parameters (i.e., $i \in Tp_{kl}$), and un-shared parameters (i.e., $i \in Tp_k - Tp_{k\cdot l}$ and $i \in Tp_l - Tp_{k\cdot l}$).

We hereafter determine and analyse Δ_{kl} from four cases, depending on whether strategy k and l use the same attack scenario (i.e., $e_k = e_l$) and whether they attack the same target (i.e., $target_k = target_l$).

Case 1: $e_k = e_l$, $target_k = target_l$.

The two strategies have the same attack scenario in case 1, which implies that their attack cost should be the same, i.e., $C_l = C_k$. Moreover, the attack targets are the same. Thus we have also the same conditional expected loss, i.e., $\widetilde{PL}_k = \widetilde{PL}_l$. Taking these analysis observations into consideration, Formula (4.16) can be simplified as:

$$\Delta_{kl} = \sum\nolimits_{j\in N} \left[\left(\prod\nolimits_{i\in Tp_{kl}} \tilde{p}_{ij}\right) \cdot \widetilde{PL}_k \cdot \left(\prod\nolimits_{i\in Tp_k - Tp_{kl}} \tilde{p}_{ij} - \prod\nolimits_{i\in Tp_l - Tp_{kl}} \tilde{p}_{ij}\right) \right] \cdot y_j \quad (4.17)$$

Recalling that only the unshared parameters are independent, which means that they can be equal to their maximal value in one strategy and be equal to their minimal value in another, we have:

$$\Delta_{kl} \geq \sum\nolimits_{j\in N} \left[\left(\prod\nolimits_{i\in Tp_{kl}} \tilde{p}_{ij}\right) \cdot \widetilde{PL}_k \cdot \left(\prod\nolimits_{i\in Tp_k - Tp_{kl}} \tilde{p}_{ij}{}^{min} - \prod\nolimits_{i\in Tp_l - Tp_{kl}} \tilde{p}_{ij}{}^{max}\right) \right] \cdot y_j$$

$$(4.18)$$

Define $\xi_{klj}^{min} = \left(\prod\nolimits_{i\in Tp_k - Tp_{kl}} \tilde{p}_{ij}{}^{min} - \prod\nolimits_{i\in Tp_l - Tp_{kl}} \tilde{p}_{ij}{}^{max}\right)$. For $\forall j \in N$ in Formula (4.18), inequality (4.19) would hold if $\xi_{klj}^{min} \leq 0$, and vice versa.

$$\left[\left(\prod_{i\in Tp_{kl}}\tilde{p}_{ij}\right)\cdot\widetilde{PL}_k\cdot\xi_{klj}^{min}\right]\cdot y_j \geq \left[\left(\prod_{i\in Tp_{kl}}\tilde{p}_{ij}^{max}\right)\cdot\widetilde{PL}_k^{max}\cdot\xi_{klj}^{min}\right]\cdot y_j \qquad (4.19)$$

Noting that Formula (4.18) is a polynomial, thus we may derive:

$$\Delta_{kl} \geq \sum_{j\in N}\left[\left(\prod_{i\in Tp_{kl}}\tilde{p}_{ij}^{\varphi}\right)\cdot\widetilde{PL}_k^{\varphi}\cdot\xi_{klj}^{min}\right]\cdot y_j = \Delta_{kl}^{min} \qquad (4.20)$$

In which:

$$\varphi = \begin{cases} max, if\ \xi_{klj}^{min} \leq 0 \\ min, otherwise \end{cases}$$

Analogously, we have:

$$\Delta_{kl}^{max} = \sum_{j\in N}\left[\left(\prod_{i\in Tp_{kl}}\tilde{p}_{ij}^{\varphi}\right)\cdot\widetilde{PL}_k^{\varphi}\cdot\xi_{klj}^{max}\right]\cdot y_j \qquad (4.21)$$

in which:

$$\xi_{ktj}^{max} = \left(\prod_{i\in Tp_k-Tp_{lk}}\tilde{p}_{ij}^{max} - \prod_{i\in Tp_l-Tp_{lk}}\tilde{p}_{ij}^{min}\right)$$

$$\varphi = \begin{cases} max, if\ \xi_{klj}^{max} \geq 0 \\ min, otherwise \end{cases}$$

Case 2: $e_k = e_l,\ target_k \neq target_l$

In this second case, the two strategies have the same attack scenario, hence their attack costs are identical, i.e., $C_l = C_k$. The attack targets are however different. Thus we have different conditional expected losses, i.e., $\widetilde{PL}_k \neq \widetilde{PL}_l$. In this case 2, Formula (4.16) can thus be reformulated as:

$$\Delta_{kl} = \sum_{j\in N}\left[\left(\prod_{i\in Tp_{kl}}\tilde{p}_{ij}\right)\cdot\left(\prod_{i\in Tp_k-Tp_{kl}}\tilde{p}_{ij}\cdot\widetilde{PL}_k - \prod_{i\in Tp_l-Tp_{kl}}\tilde{p}_{ij}\cdot\widetilde{PL}_l\right)\right]\cdot y_j$$
$$(4.22)$$

Recalling that only the unshared parameters are independent, which means that they can be equal to their maximal value in one strategy and be equal to their minimal value in another, we have:

$$\Delta_{kl} \geq \sum_{j\in N}\left[\left(\prod_{i\in Tp_{kl}}\tilde{p}_{ij}\right)\cdot\left(\prod_{i\in Tp_k-Tp_{kl}}\tilde{p}_{ij}^{min}\cdot\widetilde{PL}_k^{min} - \prod_{i\in Tp_l-Tp_{kl}}\tilde{p}_{ij}^{max}\cdot\widetilde{PL}_l^{max}\right)\right]$$
$$\times\cdot y_j$$
$$(4.23)$$

Define $\xi_{klj}^{min} = \left(\prod_{i\in Tp_k-Tp_{kl}}\tilde{p}_{ij}^{min}\cdot\widetilde{PL}_k^{min} - \prod_{i\in Tp_l-Tp_{kl}}\tilde{p}_{ij}^{max}\cdot\widetilde{PL}_l^{max}\right)$. For $\forall j \in N$ in Formula (4.23), inequality (4.24) would hold if $\xi_{klj}^{min} \leq 0$, and vice versa.

$$\left[\left(\prod_{i\in Tp_{kl}}\tilde{p}_{ij}\right)\cdot\xi_{klj}^{min}\right]\cdot y_j \geq \left[\left(\prod_{i\in Tp_{kl}}\tilde{p}_{ij}^{\ max}\right)\cdot\xi_{klj}^{min}\right]\cdot y_j \qquad (4.24)$$

Noting that Formula (4.23) is a polynomial, we obtain:

$$\Delta_{kl} \geq \sum_{j\in N}\left[\left(\prod_{i\in Tp_{kl}}\tilde{p}_{ij}^{\ \varphi}\right)\cdot\xi_{klj}^{min}\right]\cdot y_j = \Delta_{kl}^{min} \qquad (4.25)$$

in which:

$$\varphi = \begin{cases} max, \text{if } \xi_{klj}^{min} \leq 0 \\ min, \text{otherwise} \end{cases}$$

Analogously, we have:

$$\Delta_{kl}^{max} = \sum_{j\in N}\left[\left(\prod_{i\in Tp_{kl}}\tilde{p}_{ij}^{\ \varphi}\right)\cdot\xi_{klj}^{max}\right]\cdot y_j \qquad (4.26)$$

in which:

$$\xi_{klj}^{max} = \left(\prod_{i\in Tp_k-Tp_{lk}}\tilde{p}_{ij}^{\ max}\cdot\widetilde{PL}_k^{max} - \prod_{i\in Tp_l-Tp_{lk}}\tilde{p}_{ij}^{\ min}\cdot\widetilde{PL}_l^{min}\right)$$
$$\varphi = \begin{cases} max, \text{if } \xi_{klj}^{max} \geq 0 \\ min, \text{otherwise} \end{cases}$$

Case 3: $e_k \neq e_l$, $target_k = target_l$, and Case 4: $e_k \neq e_l$, $target_k \neq target_l$

The third and fourth cases are analysed together. The attacker uses different attack scenarios, thus the cost related to the attack will be different, i.e., $C_l \neq C_k$. Furthermore, the conditional expected loss would also be different since the attack scenarios are different (even for the same target, different attack scenario may result in different consequences). Thus we have $\widetilde{PL}_k \neq \widetilde{PL}_l$. Analogously to the previous two cases, we directly discuss the result of case 3 and case 4 hereafter.

$$\Delta_{kl}^{min} = \sum_{j\in N}\left[\left(\prod_{i\in Tp_{kl}}\tilde{p}_{ij}^{\ \varphi}\right)\cdot\xi_{klj}^{min}\right]\cdot y_j + C_l^{min} - C_k^{max} \qquad (4.27)$$

in which:

$$\xi_{klj}^{min} = \left(\prod_{i\in Tp_k-Tp_{kl}}\tilde{p}_{ij}^{\ min}\cdot\widetilde{PL}_k^{min} - \prod_{i\in Tp_l-Tp_{kl}}\tilde{p}_{ij}^{\ max}\cdot\widetilde{PL}_l^{max}\right)$$
$$\varphi = \begin{cases} max, \text{if } \xi_{klj}^{min} \leq 0 \\ min, \text{otherwise} \end{cases}$$

And,

$$\Delta_{kl}^{max} = \sum_{j\in N}\left[\left(\prod_{i\in Tp_{kl}}\tilde{p}_{ij}^{\ \varphi}\right)\cdot\xi_{klj}^{max}\right]\cdot y_j + C_l^{max} - C_k^{min} \qquad (4.28)$$

in which:

$$\xi_{klj}^{max} = \left(\prod_{i\in Tp_k - Tp_{kl}} \tilde{p}_{ij}^{max} \cdot \widetilde{PL}_k^{max} - \prod_{i\in Tp_l - Tp_{kl}} \tilde{p}_{ij}^{min} \cdot \widetilde{PL}_l^{min}\right)$$

$$\varphi = \begin{cases} max, if\ \xi_{klj}^{max} \geq 0 \\ min, otherwise \end{cases}$$

Proposition 4.1 Δ_{kl}^{min} and Δ_{kl}^{max} are the maximal and the minimal bounded of Δ_{kl}.

Proposition 4.2 $\Delta_{kl}^{min} = -\Delta_{lk}^{max}$.

We can remark that this second proposition can be proven considering the fact that for each of the above mentioned four cases, we have $\xi_{klj}^{min} = -\xi_{lkj}^{max}$. Hereafter, only the Δ_{kl}^{min} for each attacker strategy pair will be investigated.

Proposition 4.3 If $\Delta_{kl}^{min} > 0$, strategy k is always a better response than l, in case of the defender's committed mixed strategy y.

We can remark here that this proposition can be straightforwardly proven since $0 < \Delta_{kl}^{min} \leq \Delta_{kl} = U_a(k, :) \cdot y - U_a(l, :) \cdot y$.

Aside from the $c2$ in the IBGS, proposition 4.3 shows a new criterium to calculate the attacker's possible best responses, that is, for any attacker pure strategy l, if there exists a strategy $k \in M$ that satisfies $\Delta_{kl}^{min} > 0$, then strategy l must not be the attacker's best response to the defender's strategy y.

Δ_{kl}^{min} is a linear polynomial of y. The coefficient of y_i can be calculated as long as the interval parameters are given. Firstly, decide which case (i.e., the above mentioned four cases) the pure strategy pair (k, l) belongs to; secondly, employ the corresponding formulas to calculate the coefficient. For instance, in case 1, Formulas (4.19) and (4.20) must be used. Define a 3-dimensional coefficient matrix $\Omega^t(M^t, M^t, N)$, whose unit $\Omega^t(k, l, :)$ is the coefficient vector of y in Δ_{kl}^{min}.

4.5   Interval CPP Game Solver (ICGS)

Taking into account the parameter coupling problem in the interval CPP game, the game would be played as:

1. The defender commits a mixed strategy $y \in Y$;
2. Each type of attacker t observes y, and plays his best response $BR^t \in M^t$ to y;
3. The defender also tries to work out each type of attacker's best response. However, due to the existence of the distribution-free uncertainties on the attacker's payoffs, she is not able to work out the BR^t;
4. Instead, the defender commits a $y \in \bar{Y} \subset Y$ which results in strategy k^t being the highest low bound payoff for the attacker type t;
5. For any other attacker strategies, i.e., $\forall l \in M^t$, if $\Delta_{k^t l}^{min} > 0$, l would definitely not be the attacker's best response, while if $\Delta_{k^t l}^{min} \leq 0$, then l would possibly be the

attacker's best response. Define $PBR^t = \{l \in M^t | \Delta_{k^t l}^{min} \leq 0\}$ denoting the attacker's possible best response set;

6. The defender only knows that the attacker will play a strategy from the PBR^t, and she does not know which strategy exactly that the attacker would play;
7. The defender thus conservatively assumes that the attacker would play a strategy from the PBR^t which would minimize the defender's payoff.
8. The defender plays her optimal action.

Formula (4.29) shows a Mixed Integer Linear Programming (MILP) based algorithm, for calculating the equilibrium for the Bayesian Stackelberg CPP game with distribution-free uncertainties. Being different to the IBGS algorithm, which is general for solving any bi-matrix games with distribution-free uncertainties, the algorithm shown in Formula (4.29) considers the parameter coupling problem in the CPP game, and therefore it is only applicable for solving the CPP game with distribution-free uncertainties. For this reason, the algorithm shown in Formula (4.29) is named as "Interval CPP Game Solver" (ICGS).

In the ICGS algorithm as shown in Formula (4.29), Constraint $c9$ denotes that k^t is assumed to have the highest lower bound payoff for attacker type t (step 4) and this constraint limits the defender to play a subset of her mixed strategy (i.e., $\bar{Y} \in Y$ that satisfies constraint $c9$). Constraint $c10$ picks out the attacker's possible best response strategies. In constraint $c10$, if $\Omega^t(k^t, i, :) \cdot y > 0$, then $q_i^t = 0$; if $\Omega^t(k^t, i, :) \cdot y < 0$, then $q_i^t = 1$; otherwise q_i^t can be either 0 or 1. Comparing to step 5 in the above text, q_i^t indicates whether strategy i belongs to the PBR^t (i.e., $q_i^t = 1$) or not (i.e., $q_i^t = 0$). Constraint $c11$ together with the cost function denotes that among all the attacker's possible best responses (i.e., PBR^t), the defender conservatively thinks that the attacker strategy being worst to the defender is the actual best response strategy for the attacker (step 7 and 8). In the cost function, the defender is maximizing γ while in constraint $c11$, γ is required to be less than any $U_d^t(i, :) \cdot y$ (which is the defender's payoff in case that the attacker plays strategy i as a response to the defender's committed strategy y) if $q_i^t = 1$ (which means that strategy i is in the attacker's possible best response strategy set). Notice that in constraint $c11$, if $q_i^t = 0$, then the constraint is not activated (Γ is a big constant number and therefore the left-hand side of the inequality is always greater than the right-hand side). Constraints $c12$ and $c13$ express that q and y are binary variables and a mixed strategy, respectively. $c12$ also mentions $q_{k^t}^t = 1$. The reason is that: (i) obviously we have that $\Omega^t(k^t, k^t, :) \cdot y = 0$; (ii) according to constraint $c10$, $q_{k^t}^t$ can be either 0 or 1, which means that strategy k^t can be either in the attacker's best response strategy set or not; (iii) however, k^t is the strategy that has the highest attacker lower bound payoff (i.e., $R^t = \underline{U}_a^t(k^t, :) \cdot y = \max_{i \in M^t} \{\underline{U}_a^t(i, :) \cdot y\}$) and the attacker's upper bound payoff by playing strategy k^t would be no less than the lower bound payoff (i.e., $\bar{U}_a^t(k^t, :) \cdot y \geq_{U_{a^t}(k^t,:) \cdot y = R^t}$), thus strategy k^t should be included in the PBR^t;

(iv) we manually set $q_{k^t}^t = 1$.

$$\max_{q,\gamma,y} \sum_{t \in TL} \rho^t \gamma^t$$

$$s.t. \begin{cases} c9. \; \underline{U}_a^t(i,:) \cdot y \leq \underline{U}_a^t(k^t,:) \cdot y, \; \forall i \in M^t \\ c10. \; -q_i^t \cdot \Gamma \leq \Omega^t(k^t,i,:) \cdot y \leq (1-q_i^t) \cdot \Gamma, \; \forall i \in M^t \\ c11. \; \Gamma \cdot (1-q_i^t) + U_d^t(i,:) \cdot y \geq \gamma^t, \; \forall i \in M^t \\ c12. \; q_i^t \in \{0,1\}, q_{k^t}^t = 1 \\ c13. \; \sum y = 1, y_i \in [0,1] \end{cases} \quad (4.29)$$

The algorithm should be implemented to each combination of $k^t \in M^t$, obtaining the optimal payoff for the defender $H(k^1, k^2, \ldots, k^{|TL|})$ and the corresponding optimal strategy for the defender $\tilde{y}(k^1, k^2, \ldots, k^{|TL|})$. Finally, choose the maximal H and its corresponding $\tilde{y}$ as the defender's optimal payoff and optimal solution.

Proposition 4.4 Defender's equilibrium payoff from the ICGS is higher than or equal to her equilibrium payoff from the IBGS.

Proof $\forall y \in Y$, without loss of generality, assume that $\underline{U}_a^t(\pi^t,:) \cdot y \geq \underline{U}_a^t(i,:) \cdot y$, for all $i \in M^t$.

In the IBGS, from $c1$ we have $R^t = \underline{U}_a^t(\pi^t,:) \cdot y$. From $c2$ we know that $\forall i \in M^t$, if $\bar{U}_a^t(i,:) \cdot y < R^t$, then $q_i^t = 0$, which means that the strategy i will definitely not be the attacker's best response. Define $E_B^t = \{i \in M^t - \{\pi^t\} | \bar{U}_a^t(i,:) \cdot y < R^t\}$. According to $c3$, we have $\gamma_B^t = min_{i \in M^t - E_B^t} \{U_d^t(i,:) \cdot y\}$.

In the ICGS, from $c9$ we have $k^t = \pi^t$. From $c10$ we know that $\forall i \in M^t$, if $\Omega^t(k^t, i,:) \cdot y > 0$, then $q_i^t = 0$, which means that strategy k^t is always a better response than strategy i, or, strategy i will definitely not be the attacker's best response. Define $E_C^t = \{i \in M^t - \{\pi^t\} | \Omega^t(k^t, i,:) \cdot y > 0\}$. According to $c11$, we have that $\gamma_C^t = min_{i \in M^t - E_C^t} \{U_d^t(i,:) \cdot y\}$.

Now we are going to prove that $E_B^t \subseteq E_C^t$. $\forall i \in E_B^t$, we have $0 < \underline{U}_a^t(\pi^t,:) \cdot y - \bar{U}_a^t(i,:) \cdot y \leq min\{U_a^t(\pi^t,:) \cdot y - U_a^t(i,:) \cdot y\} = \Omega^t(k^t, i,:) \cdot y$, thereby $i \in E_C^t$.

Since $E_B^t \subseteq E_C^t$, we have $\gamma_B^t \leq \gamma_C^t$, thus $\sum_{t \in TL} \rho^t \gamma_B^t \leq \sum_{t \in TL} \rho^t \gamma_C^t$. $\square$

4.6 Conclusion

In this chapter, the interval CPP game is defined and two algorithms are proposed. In the interval CPP game, the defender knows an interval in which the attacker's parameters will be located, however she does not know what the exact values of the parameters are and how the parameters are distributed in the interval. The defender thus plays conservatively to make the best use of her knowledge. The two proposed algorithms are both based on Mixed Integer Linear Programming techniques. The interval bi-matrix game solver (IBGS) is applicable to any bi-matrix games with such distribution-free uncertainties, while the interval CPP game solver

(ICGS), is proposed only for solving CPP games with interval inputs. A theoretic study shows that the ICGS algorithm can bring the defender a higher or an equal payoff when compared to the IBGS algorithm.

In industrial practice, it is difficult to obtain the exact numbers of the parameters that a CPP game needs. As also illustrated in Sect. 3.4.1, input data from conventional security risk assessment methods is always related to an interval, see for instance, Table 3.3, 3.4, and 3.5. With the work in this chapter, the CPP game model is able to directly deal with the data obtained from the application of those conventional security risk assessment methods.

References

1. Zhang L, Reniers G, Qiu X. Playing chemical plant protection game with distribution-free uncertainties. Reliab Eng Syst Saf. 2017.
2. Dantzig G. Linear programming and extensions. Princeton: Princeton University Press; 2016.
3. Ben-Tal A, Nemirovski A. Robust solutions of uncertain linear programs. Oper Res Lett. 1999;25(1):1–13.
4. Kiekintveld C, Marecki J, Tambe M, editors. Approximation methods for infinite Bayesian Stackelberg games: modeling distributional payoff uncertainty. In: The 10th international conference on autonomous agents and multiagent systems-volume 3. International Foundation for Autonomous Agents and Multiagent Systems; 2011.
5. Kiekintveld C, Islam T, Kreinovich V, editors. Security games with interval uncertainty. In: Proceedings of the 2013 international conference on Autonomous agents and multi-agent systems; 2013: International Foundation for Autonomous Agents and Multiagent Systems; 2013.
6. Nikoofal ME, Zhuang J. Robust allocation of a defensive budget considering an attacker's private information. Risk Anal. 2012;32(5):930–43.
7. Pita J, Jain M, Tambe M, Ordóñez F, Kraus S. Robust solutions to Stackelberg games: addressing bounded rationality and limited observations in human cognition. Artif Intell. 2010;174 (15):1142–71.

Chapter 5
Single Plant Protection: Playing the Chemical Plant Protection Game Involving Attackers with Bounded Rationality

In this chapter, we model attackers with bounded rationality in the Chemical Plant Protection game. Three different behaviour models of attackers are investigated, namely, the epsilon-optimal attacker, the monotonic-optimal attacker, and the Mini-Max attacker. All these attacker models are integrated to the Stackelberg CPP game, which means that the defender moves first, and the attackers follow. Furthermore, the monotonic-optimal attacker is investigated in the Interval CPP game with only one type of attacker, and a game solution named Monotoic MaxiMin Solution for the Interval CPP game (MoSICP) is defined [1]. The MoSICP solution incorporates both bounded rational attackers and distribution-free uncertainties into the CPP game. The epsilon-optimal attacker model, being related to the defender's distribution-free uncertainties, and the MiniMax attacker model, being the most conservative model, are therefore investigated in the Bayesian Stackelberg CPP game framework, instead of in the Interval CPP game framework. The defender is still assumed to behave rationally to maximize her payoff.

5.1 Motivation

The chemical plant protection (CPP) game, proposed in Chap. 3, is able to model intelligent interactions between the (industrial) defender and potential so-called adversaries (or in other words, human attackers). Moreover, by extending the CPP game to a Bayesian game, multiple types of attackers can be modelled in the game. The interval CPP game, proposed in Chap. 4, fills the gap between the difficult requirement of a lot of quantitative data for the CPP game and the difficulties of

---

This chapter is mainly based on the paper of Zhang et al. [1]

© Springer International Publishing AG, part of Springer Nature 2018
L. Zhang, G. Reniers, *Game Theory for Managing Security in Chemical Industrial Areas*, Advanced Sciences and Technologies for Security Applications,
https://doi.org/10.1007/978-3-319-92618-6_5

obtaining these input numbers. Nonetheless, all previous models assume rational players in the game.

However, the players of the CPP game (i.e., the defender and the adversaries) are not always rational players, consider for instance terrorist attackers. Guikema [2] points out that the rationality assumption brings modelling convenience to the modellers and that it is also a common assumption in a wide variety of fields. However, Guikema [2] further indicates that spontaneous attackers are not behaving to maximize their subjective expected utilities while in case of a premediated attacker, it is difficult to quantify the attacker's emotional dimension (e.g., honor). Therefore, game theoretic models must be extended to be able to deal with such 'bounded-rational' attackers.

Many researchers study games with bounded-rational players. Among others, in the security game domain, several representative attacker models are:

1. the epsilon-optimal attacker [3], as also explained in Sect. 5.2 of this book.
2. the quantal response equilibrium (QRE) [4, 5], which is only defined for games with discrete strategies and the probabilities that each pure strategy would be played can be calculated by Formula (5.1), in which x_i denotes the probability that pure strategy i would be played, $EXP(\cdot)$ represents the exponential function, λ is a constant real number, A is the payoff matrix and $e_i \cdot A \cdot y$ is the player's payoff by responding strategy i to his opponent's strategy y, m is the number of pure strategies.

$$x_i = \frac{EXP(\lambda \cdot e_i \cdot A \cdot y)}{\sum_{k=1,2,...,m} EXP(\lambda \cdot e_k \cdot A \cdot y)} \tag{5.1}$$

3. the *level* $-$ *k* thinking model [6], which assumes that a level k ($k = 1, 2, \ldots$) player is playing the game optimally presuming his opponent is a level $(k - 1)$ player and the level 0 player is a non-strategic player who randomly chooses his strategies.
4. the monotonic optimal attacker [7, 8], as explained in Sect. 5.3 in this book.
5. the MiniMax attacker, as explained in Sect. 5.4 in this book.

The quantal response equilibrium and the *level* $-$ *k* thinking model strictly rely on their modelling parameters, namely the λ for the QRE and the k for the *level* $-$ *k* model. These parameters are difficult, if not impossible, to be obtained. Tambe and his co-authors employed both simulation games and machine learning techniques for estimating the λ for the QRE [9]. There are also real experiments, for instance, the "beauty contest games" [10], for evaluating how people behave (i.e., what the value of k should be) in the *level* $-$ *k* thinking framework. However, on the one hand these parameters can vary between people and on the other hand a terrorist may be thinking completely different from ordinary people, and thus, the QRE and the *level − k* thinking model are not investigated in this book for the Chemical Plant Protection game.

5.2   Epsilon-Optimal Attacker

5.2.1   Definition of an 'Epsilon-Optimal Attacker'

Definition 5.1 An attacker is called an epsilon-optimal attacker if he would play any strategy from a possible strategy set *PS* as defined in Formula (5.2), under a condition that the defender commits a mixed strategy y.

$$PS = \{s_a \in S_a | BRP - U_a(s_a, :) \cdot y \le \varepsilon\} \tag{5.2}$$

In Formula (5.2), *BRP* represents the attacker's best response payoff to the defender's strategy y, i.e., $BRP = \max_{s_a \in S_a} U_a(s_a, :) \cdot y$; ε is a user-defined constant real number.

The definition implies (and reveals) that an epsilon-optimal attacker would not always play his best response strategy, instead, he may deviate to any strategy having a close payoff to his best response payoff.

The 'epsilon-optimal attacker' concept is investigated in this chapter because, on the one hand, such concept represents a typical human behaviour rule. Human beings are not machines and when they make decisions or do calculations, small errors are generally ignored. In daily lives, people do not even carry out numerical calculations, and they make decisions depending just on a fuzzy intuitive. For instance, in a plant with multiple dangerous oil tanks, the defender and a professional (knowledgeable) attacker can perform certain calculations and, by doing so, know which tank will lead to the most severe consequences, and determine a priority list of tanks to attack. However, a non-professional adversary not carrying out any calculations, may have identical interests with respect to the different oil tanks, since all these oil tanks have a certain dangerousness level according to intuition.

On the other hand, the epsilon-optimal attacker model may also reflect the defender's uncertainties on the attacker's payoff. The attacker's best response payoff *BRP* is calculated based on his payoff matrix. In reality, the defender may not know the exact numbers of the attacker's payoff matrix, thus there might be errors on the defender's estimation on the attacker's *BRP*. Therefore, the defender sets a tolerance (i.e., ε) to increase the robustness of her decision. In Sect. 5.2.3, we will see that the algorithm for solving the chemical plant protection game played by an epsilon-optimal attacker is actually similar to the algorithm IBGS that we developed for solving general bi-matrix games with distribution-free uncertainties.

5.2.2   Game Modelling of the 'Epsilon-Optimal Attacker'

In the definition of the Stackelberg equilibrium for the CPP game (see Formulas (3.19) and (3.20), in Sect. 3.3.2), the defender first commits a mixed strategy y, then the attacker plays his best response, and the defender can also work out the attacker's best response, thus the defender plays accordingly. In the CPP game played by a rational defender and an epsilon-optimal attacker, the procedure can be illustrated as follows:

1. the defender commits to a mixed strategy y;
2. the attacker calculates his best response strategy and the corresponding best response payoff (BRP);
3. the defender can also work out the attacker's BRP, and she furthermore calculates the attacker's possible strategies set PS, according to Formula (5.2);
4. the defender conservatively thinks that the attacker would play a strategy $s_a \in PS$ which minimizes her payoff;
5. based on procedures (1)–(4), the defender plays optimally.

Steps 3 and 4 are different to the procedure of calculating the Stackelberg equilibrium. In step 3, the attacker is assumed bounded-rational, thus the defender can only work out a possible strategies set, instead of knowing that the attacker will definitely play his best response strategy. In step 4, the defender does not know which strategy the attacker would use, thus she chooses to play conservatively, presuming that among all possible strategies in the PS, the worst one to her will be the attacker's choice.

5.2.3   Solving the CPP Game with 'Epsilon-Optimal Attackers'

As we discussed in 5.2.1, an epsilon-optimal attacker model can also reflect the defender's uncertainties on the attacker's payoff. Therefore, an algorithm as shown in Formula (5.3) is proposed to solve the CPP game with epsilon-optimal attackers. The algorithm deviates from the IBGS, which is developed in Sect. 4.3 for solving games with distribution-free uncertainties. Only the constraints $c1$ and $c2$ in Formula (5.3) are different from the constraints in the IBGS, while other constraints as well as the cost functions and the variables are the same. For this reason, we only explain constraints $c1$ and $c2$ in this algorithm and further refer to the explanation of the other elements of the algorithm given in Sect. 4.3 on the IBGS algorithm.

$$\max_{q,h,\gamma,y,R} \sum_{t\in TL} p^t \gamma^t$$

$$s.t. \begin{cases} c1. \ 0 \le R^t - U_a^t(i,:)\cdot y \le (1 - h_i^t)\cdot \Gamma, \ \forall i\in M^t \\ c2. \ (q_i^t - 1)\cdot \Gamma \le U_a^t(i,:)\cdot y - R^t + \varepsilon^t \le q_i^t \cdot \Gamma, \ \forall i\in M^t \\ c3. \ \Gamma\cdot (1 - q_i^t) + U_a^t(i,:)\cdot y \ge \gamma^t, \ \forall i\in M^t \\ c4. \ q_i^t \ge h_i^t, \ \forall i\in M^t \\ c5. \ q_i^t, h_i^t \in \{0,1\} \\ c6. \ \sum h^t = 1 \\ c7. \ \sum y = 1, y_i \in [0,1] \\ c8. \ R^t, \gamma^t \in R \end{cases} \qquad (5.3)$$

In Formula (5.3), $U_a^t(i,:)$ is the ith row of the attacker t's payoff matrix; ε^t is the user-defined tolerance of attacker t. Other notations are the same as explained in Sect. 4.3 below the IBGS algorithm. Constraint $c1$ calculates the attacker's best response payoff R^t. Note that in $c1$, if $h_i^t = 1$, then $R^t = U_a^t(i,:)\cdot y$, while if $h_i^t = 0$, then $R^t \ge U_a^t(i,:)\cdot y$. Therefore, $R^t = \max_{i\in M^t} U_a^t(i,:)\cdot y$ and $h_i^t = 1$ means that strategy i is the attacker t's best response strategy to the defender's committed strategy y. Constraint $c2$ defines the attacker's possible response strategy set PS. Note that in $c2$, if $R^t - U_a^t(i,:) < \varepsilon^t$, then $q_i^t = 1$, if $R^t - U_a^t(i,:) > \varepsilon^t$, then $q_i^t = 0$, while if $R^t - U_a^t(i,:) = \varepsilon^t$, then q_i^t can be either 0 or 1. Therefore, $q_i^t = 1$ indicates that strategy i belongs to the PS, for attacker type t (see the definition of PS in Formula (5.2)).

Similar to the IBGS algorithm, the above algorithm calculates the defender's conservative payoff, knowing that she is playing the game with epsilon-optimal attackers.

5.3 Monotonic Optimal Attacker

5.3.1 Definition of a 'Monotonic Optimal Attacker'

Definition 5.2 An attacker is called a monotonic optimal attacker if he plays a mixed strategy $x \in Q(y)$ as defined in Formula (5.4), under the condition that the defender commits a mixed strategy y.

$$Q(y) = \{x\in X | \forall (i,j)\in E(y), x_i \ge x_j\} \qquad (5.4)$$

in which,

$$E(y) = \{(i,j) | U_a(i,:)\cdot y \ge U_a(j,:)\cdot y, \forall i,j = 1,2,\ldots,m; i\ne j\} \qquad (5.5)$$

for a CPP game or a general bi-matrix game, and,

$$E(y) = \left\{ (i,j) | \Delta_{ij}^{min} \geq 0, \forall i,j = 1,2,\ldots,m; i \neq j \right\} \qquad (5.6)$$

for an interval CPP game.

$E(y)$ is a set of attacker's pure strategy pairs of which the former strategy would bring higher payoff to the attacker than the latter strategy. $Q(y)$ is a subset of the attacker's mixed strategy space, satisfying that if a pure strategy pair $(i,j) \in E(y)$, then the probability that the attacker would play strategy i will be higher than the probability that he plays strategy j.

The definitions of $Q(y)$ and $E(y)$ reveal the key property of the monotonic optimal attacker model, that is, *for a committed defender's strategy y, the attacker is assumed to be more likely to respond with a pure strategy with higher payoff than to respond with a pure strategy that has a lower payoff.* For instance, knowing the defender's strategy y, if the attacker would have a payoff u_1 by playing strategy s_{a1} and a payoff u_2 by playing strategy s_{a2}, then if $u_1 \geq u_2$, the attacker would be more likely to play strategy s_{a1} than to play s_{a2} (but s_{a2} still has a probability of being played), and vice versa.

Comparing to the epsilon-optimal attacker model, the quantal response attacker model, and the *level − k* thinking attacker model, the assumption of the monotonic optimal attacker is quite relaxed. In the epsilon-optimal attacker model, a user-defined tolerance (i.e., ε) is required. However, this tolerance can be difficult to estimate. In the quantal response attacker model, a model-specific parameter λ is required, and the attacker's behaviour critically depends on that parameter, see Formula (5.1). In the *level − k* thinking model, it is difficult to decide which level that the attacker is situated in, thus the defender would not be able to operate optimally. The monotonic optimal attacker model, however, does not depend on any extra parameter.

5.3.2   Game Modelling of the 'Monotonic Optimal Attacker'

A **Mo**notonic MaxiMin **S**olution for the **I**nterval **CP**P game (MoSICP) is

$$\text{argmax}_{y \in Y} \min_{x \in Q(y)} x^T \cdot U_d \cdot y \qquad (5.7)$$

In which $Q(y)$ is as defined in Formulas (5.4) and (5.6).

The definition of the MoSICP is different to the Monotonic MaxiMin Solution (MMS) defined in Jiang et al. [7], of which the $Q(y)$ is defined by Formulas (5.4) and (5.5). Figure 5.1 illustrates the differences of the definition of $E(y)$ in the MoSICP and in the MMS. The adversary's pure strategies are shown on the X axis, while the Y axis represents the defender's estimation of the attacker's payoff. The defender is assumed to play a mixed strategy, and the attacker has seven different pure strategies to respond. The attacker's payoff by responding different strategies, without considering distribution-free uncertainties, are represented by red dots in the figure. The

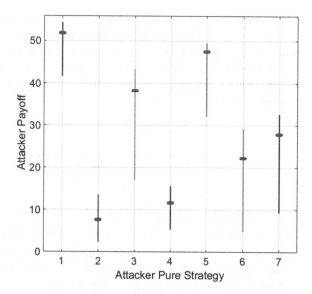

Fig. 5.1 Attacker's payoff by responding different pure strategies to y

defender's estimation of the range of the attacker's payoffs, when there are distribution-free uncertainties, are denoted by the vertical lines. Furthermore, in this illustrative figure, the parameter coupling problem descript in Chap. 4 is excluded. Therefore, Δ_{kl}^{min} would be equal to the lowest point of vertical line k minus the highest point of vertical line l.

Formula (5.8) shows the $E(y)$ for the MMS according to Formulas (5.4) and (5.5). For example, Fig. 5.1 shows that the attacker's payoff by responding strategy 1 is higher than by responding strategy 2, thus $(1, 2) \in E(y)$ in Formula (5.8). Other strategy pairs can be explained analogously. Formula (5.9) demonstrates the $E(y)$ for the MoSICP, according to Formulas (5.4) and (5.6). For example, $(3, 4) \in E(y)$ in Formula (5.9), since in Fig. 5.1, the lowest point of the third vertical line is higher than the highest point of the fourth vertical line.

$$\text{For the } MMS, E(y) = \left\{ \begin{array}{l} (1,2), (1,3), (1,4), (1,5), (1,6), (1,7), (3,2), \\ (3,4), (3,6), (3,7), (4,2), (5,2), (5,3), (5,4), \\ (5,6), (5,7), (6,2), (6,4), (7,2), (7,4), (7,6) \end{array} \right\}. \quad (5.8)$$

For the MoSICP, $E(y) = \{(1,2), (1,4), (1,6), (1,7), (3,2), (3,4), (5,2), (5,4), (5,6)\}.$

$$(5.9)$$

5.3.3   Calculating the MoSICP

The definition of the MoSICP contains both the maximal and the minimal optimization problem, and the inner minimal problem (i.e., $\min_{x \in Q(y)} x^T \cdot U_d \cdot y$) involves a Q

(y), thus it is not a standard linear minimal problem. Jiang et al. [7] proposed an approach for transforming this problem into a standard linear problem by using duality. In this book, we mainly follow the idea from Jiang et al. [7]

For a given y, the inner optimization problem (i.e., $\min_{x \in Q(y)} x^T \cdot U_d \cdot y$) in the MoSICP can be written as Formula (5.10), of which the constraints demonstrate the definition of $Q(y)$.

$$
\min_x x^T \cdot (U_d \cdot y)
$$
$$
s.t. \begin{cases} x_i \geq x_j, \forall (i, j) \in E(y) \\ x_i \geq 0 \\ \sum x = 1 \end{cases} \tag{5.10}
$$

For any attacker strategy pair $(i, j) \in E(y)$, the first constraint in Formula (5.10) can be written as $\left(e_i^m - e_j^m \right) \cdot x \geq 0$, while for any strategy pair $(i, j) \notin E(y)$, it can be written as $0_{1 \times m} \cdot x \geq 0$. e_i^m is a row vector with a length of m. The ith entry of e_i^m is 1, and other entries are 0. For instance, $e_3^4 = [0\ 0\ 1\ 0]$. $0_{1 \times m}$ is a row zero vector with a length of m. With this definition, Formula (5.10) can further be formulated as:

$$
\min_x x^T \cdot (U_d \cdot y)
$$
$$
s.t. \begin{cases} W \cdot x \geq 0 \\ x_i \geq 0 \\ \sum x = 1 \end{cases} \tag{5.11}
$$

In which W is an $m(m - 1) \times m$ matrix, whose ij^{th} row is $e_i^m - e_j^m$ if $(i, j) \in E(y)$ and is $0_{1 \times m}$ otherwise.

With a given y, the $E(y)$ is determined, thus Formula (5.11) is a standard linear programming problem. By applying the duality theory on Formula (5.11), we obtain:

$$
\max_{\theta, t} t
$$
$$
s.t. \begin{cases} \theta_i \geq 0, i = 1, 2, \ldots, M \\ W^T \cdot \theta + t \leq U_d \cdot y \end{cases} \tag{5.12}
$$

In which θ and t are the dual variables and $M = m(m - 1)$. Note that Formula (5.12) works only for a determined $E(y)$, and different y may result in different $E(y)$. Define a binary variable $z_{ij} \in \{0, 1\}$, setting $z_{ij} = 1$ if the attacker strategy pair $(i, j) \in E(y)$ and $z_{ij} = 0$ otherwise. Combining the definition of $E(y)$ (Formula (5.6)), the definition of z_{ij} can also be expressed as:

$$
\Delta_{ij}^{min} + \Gamma \cdot \left(1 - z_{ij} \right) \geq 0, \tag{5.13}
$$

and,

$$\Delta_{ij}^{min} - \Gamma \cdot z_{ij} < 0. \tag{5.14}$$

In these 2 inequalities, $\Delta_{ij}^{min} \geq 0 \Leftrightarrow z_{ij} = 1$, and $\Delta_{ij}^{min} < 0 \Leftrightarrow z_{ij} = 0$.

With the support of z_{ij}, we may rewrite the ij^{th} row of matrix W as $z_{ij} \cdot \left(e_i^m - e_j^m \right)$. Moreover, the second constraint in Formula (5.12) can be rewritten as:

$$\sum_{i=1, j=1, i \neq j}^{i=m, j=m} \bar{w}_{ij,k} \cdot \theta_{ij} \cdot z_{ij} + t \leq U_d(k,:) \cdot y, k = 1, 2, \ldots, m \tag{5.15}$$

in which $\bar{w}_{ij,k}$ is the (ij, k) entry of a matrix whose ij^{th} row is $e_i^m - e_j^m$.

Formula (5.15) is not a standard linear constraint either, due to the existence of the multiplications of variables (i.e., $\theta_{ij} \cdot z_{ij}$). We further define $\omega_{ij} = \theta_{ij} \cdot z_{ij}$, such that we obtain a mixed integer linear programming (MILP) based algorithm, for calculating the MoSICP, as shown in Formula (5.16).

$$
\begin{aligned}
&\max_{y, \omega, t, z} t \\
&s.t. \begin{cases}
c1. & \displaystyle\sum_{i=1, j=1, i \neq j}^{i=m, j=m} \bar{w}_{ij,k} \cdot \omega_{ij} + t \leq U_d(k,:) \cdot y, \quad \forall k \in M \\
c2. & 0 \leq \omega_{ij} \leq N \cdot z_{ij}, \quad \forall i, j \in M, i \neq j \\
c3. & \Delta_{ij}^{min} + N \cdot (1 - z_{ij}) \geq 0, \quad \forall i, j \in M, i \neq j \\
c4. & \Delta_{ij}^{min} - N \cdot z_{ij} < 0, \quad \forall i, j \in M, i \neq j \\
c5. & (1 - z_{ij}) + (1 - z_{jh}) + z_{ih} \geq 1, \quad \forall i, j, h \in M, i \neq j \neq h \\
c6. & z_{ij} \in \{0, 1\}, y \in Y, \quad \forall i, j \in M, i \neq j
\end{cases}
\end{aligned} \tag{5.16}
$$

In the algorithm, constraint $c1$ directly refers to Formula (5.15) (and thus it refers to the second constraint in Formula (5.12)); in constraint $c2$, if $z_{ij} = 0$, then $\omega_{ij} = \theta_{ij} \cdot 0 = 0$, if $z_{ij} = 1$, then $\omega_{ij} = \theta_{ij} \cdot 1 \geq 0$, therefore, $c2$ refers to the first constraint in Formula (5.12); constraints $c3$, $c4$ reflect the definition of z_{ij}, as also shown in Formulas (5.13) and (5.14); constraint $c5$ represents that if (i,j) and (j,h) belong to $E(y)$ then (i,h) must also belong to $E(y)$.

Proposition 6.1 According to the definition of MoSICP, if (i,j) and (j,h) belong to $E(y)$ then (i,h) must also belong to $E(y)$, that is to say, $\forall i, j, h \in M, i \neq j \neq h$, then $z_{ij} = 1$ and $z_{jh} = 1$ imply that $z_{ih} = 1$.

Proof If (i,j) and (j,h) belong to $E(y)$, then we have that $\Delta_{ij}^{min} \geq 0$ and $\Delta_{jh}^{min} \geq 0$ (see Formula (5.6)). Furthermore, we have that $0 \leq \Delta_{ij}^{min} + \Delta_{jh}^{min} = \min\{U_a(i,:) \cdot y - U_a(j,:) \cdot y\} + \min\{U_a(j,:) \cdot y - U_a(h,:) \cdot y\} \leq \min\{U_a(i,:) \cdot y - U_a(j,:) \cdot y + U_a(j,:) \cdot y - U_a(h,:) \cdot y\} = \Delta_{ih}^{min}$. Therefore, $\Delta_{ih}^{min} \geq 0$ and thus (i,h) also belongs to $E(y)$.

Constraint $c4$ is a strict inequality, which cannot be processed by typical MILP solvers. When we implement the algorithm, $c4$ is typed in as $\Delta_{ij}^{min} - \Gamma \cdot z_{ij} + \sigma \leq 0$ in which σ is a small constant real number. The usage of σ may reduce the defender's expect optimal payoff by excluding a sub-set of Y. Furthermore, the sub-set can be worked out as $\tilde{Y} = \left\{ y \in Y \mid -\sigma < \Delta_{ij}^{min} < 0, \forall i,j \right\}$. By fixing σ sufficiently small, $\tilde{Y}$ can be reasonably bounded.

When the defender's optimal strategy y is obtained, according to Formulas (5.4) and (5.6), sets $E(y)$, $Q(y)$ can be constructed. By solving the linear program $\min_{x \in Q(y)} x^T \cdot (U_d \cdot y)$, the attacker's response strategy can be calculated.

5.4   MiniMax Attacker

5.4.1   Definition of a 'MiniMax Attacker'

Definition 5.3 a MiniMax attacker is an attacker who plays totally converse to the defender's interest. That is, knowing a defender's committed strategy y, a MiniMax attacker would play a strategy $k = argmin_{s_a \in S_a} U_d(s_a, :) \cdot y$.

MiniMax attackers can be treated as totally irrational players, since they do not care about their own payoffs at all, instead, they look at minimizing the defender's payoff.

5.4.2   Game Modelling of the 'MiniMax Attacker'

In a MiniMax attacker case, the adversary's payoff is obviously no longer needed for the game modelling. The defender knows that whatever strategy she plays, the attacker aims to minimize her payoff. Therefore, the defender can play optimally as:

$$\tilde{y} = argmax_{y \in Y} \left\{ min_{x \in X} x^T \cdot U_d \cdot y \right\} \tag{5.17}$$

5.4.3   Solving the CPP Game with 'MiniMax Attackers'

Solving the CPP game with 'MiniMax attackers' is equivalent to solve a zero-sum game. Formula (5.18) shows an algorithm for calculating the defender's optimal strategy and the corresponding payoff.

$$\max_{y,\gamma} \sum_{t \in TL} p^t \cdot \gamma^t$$
$$s.t. \begin{cases} c1 : U_d^t(i,:) \cdot y \geq \gamma^t, \forall i \in M^t, t \in TL \\ c2 : y \in Y \end{cases} \qquad (5.18)$$

The first constraint shows that if the defender plays a mixed strategy y, then the attacker would play a pure strategy $i \in M$ that minimizes the defender's payoff. The cost function aims to maximize the defender's minimal payoff (result in the attacker's choice). This algorithm is developed based on an important property of the attacker's choice: knowing y, the attacker's choice would be a pure strategy, instead of being a mixed strategy.

5.5 Conclusion

Boundedly rational attackers in the chemical plant protection game are modelled in this chapter. The 'Epsilon-optimal attacker', the 'Monotonic optimal attacker', and the 'MiniMax attacker' are defined and integrated into the CPP game. Furthermore, algorithms for solving CPP games with these boundedly rational attackers are developed.

References

1. Zhang L, Reniers G, Chen B, Qiu X. A chemical plant protection game incorporating boundedly raitonal attackers and distribution-free uncertainties. Submitted to Process Safety and Environmental Protection; 2018.
2. Guikema SD. Game theory models of intelligent actors in reliability analysis: an overview of the state of the art. In: Game theoretic risk analysis of security threats. New York: Springer; 2009. p. 13–31.
3. Pita J, Jain M, Tambe M, Ordóñez F, Kraus S. Robust solutions to Stackelberg games: addressing bounded rationality and limited observations in human cognition. Artif Intell. 2010;174(15):1142–71.
4. McKelvey RD, Palfrey TR. Quantal response equilibria for normal form games. 1993.
5. Yang R, Ordonez F, Tambe M, editors. Computing optimal strategy against quantal response in security games. In: Proceedings of the 11th international conference on autonomous agents and multiagent systems-volume 2. International Foundation for Autonomous Agents and Multiagent Systems; 2012.
6. Rothschild C, McLay L, Guikema S. Adversarial risk analysis with incomplete information: a level-k approach. Risk Anal. 2012;32(7):1219–31.
7. Jiang AX, Nguyen TH, Tambe M, Procaccia AD, editors. Monotonic maximin: A robust stackelberg solution against boundedly rational followers. In: International conference on decision and game theory for security. Springer; 2013.

8. Nguyen TH, Jiang AX, Tambe M, editors. Stop the compartmentalization: unified robust algorithms for handling uncertainties in security games. In: Proceedings of the 2014 international conference on autonomous agents and multi-agent systems. International Foundation for Autonomous Agents and Multiagent Systems; 2014.
9. Nguyen TH, Yang R, Azaria A, Kraus S, Tambe M, editors. Analyzing the effectiveness of adversary modeling in security games. New York: AAAI; 2013.
10. Nagel R. Unraveling in guessing games: an experimental study. Am Econ Rev. 1995;85 (5):1313–26.

Chapter 6
Multi-plant Protection: A Game-Theoretical Model for Improving Chemical Clusters Patrolling

6.1 Introduction

Due to economies of scale and all kinds of collaboration benefits, chemical plants are usually geographically clustered, forming chemical industrial parks or so-called 'chemical clusters'. Some examples of such clusters are the Antwerp port chemical cluster in Belgium, the Rotterdam port chemical cluster in the Netherlands, the Houston chemical cluster in the US, or the Tianjin chemical cluster in China. Besides fixed security countermeasures within every plant, the patrolling of security guards is also scheduled, for securing these chemical facilities at different points and times, e.g. at night. The patrolling can either be single-plant oriented, which can be completely scheduled by the plant itself, or it can be multiple-plants oriented, which should be scheduled by an institute at a higher level than the single-plant level, for instance a multiple plant council (MPC) [1]. Both types of patrolling have a drawback of not being able to deal with intelligent attackers. Some patrollers follow a fixed patrolling route, and in this case the adversary is able to predict the patroller's position at a certain time. Other patrollers purely randomize their patrolling, without taking into consideration the hazardousness level that each installation/facility/plant holds, and if this is the case, the adversary may focus to attack the most dangerous installations/facilities/plants since all installations/facilities/plants are equally patrolled.

As we have explained before in this book, game theory is a mathematics-based methodology for modelling intelligent interactions between defenders and potential attackers. However, no research has been done thus far for employing a game theoretic model to optimize patrolling in chemical industrial parks. Nonetheless, game theory has been introduced for improving patrol scheduling in some other domains. Shieh et al. [2] proposed a game theoretic model for optimizing patrolling

---

This chapter is mainly based on the paper of Zhang and Reniers [6]

© Springer International Publishing AG, part of Springer Nature 2018
L. Zhang, G. Reniers, *Game Theory for Managing Security in Chemical Industrial Areas*, Advanced Sciences and Technologies for Security Applications,
https://doi.org/10.1007/978-3-319-92618-6_6

of protecting ferries in the Boston port. Fang et al. [3] developed the so-called green security game for scheduling patrolling for the conservation of wild animals. Amirali et al. [4] introduced a model based on game theory to better patrol pipelines. Alpern et al. [5] systematically and theoretically studied the patrolling problem in a graph.

This chapter proposes a Chemical Cluster Patrolling (CCP) game [6] answering the question how to optimally randomize patrolling in a chemical cluster, in a way that it is better secured, by using a game theoretical approach. The remainder of this chapter is organized as follows: Sect. 6.2 briefly introduces how patrolling is organized in chemical clusters. Sect. 6.3 proposes the chemical cluster patrolling game and solutions for solving the CCP game are discussed in Sect. 6.4. Conclusions are drawn in Sect. 6.5. Furthermore, a case study of the CCP game is given in Sect. 7.2.

6.2   Patrolling in Chemical Clusters

6.2.1   A Brief Patrolling Scenario Within a Chemical Cluster

The patrolling scenario is assumed to be the following. A patroller team (e.g. consisting of two guards) drives a car randomly, patrolling in each of the plants. In each plant, the team drives into the plant and conducts a patrolling task and/or some other security related actions. Besides each plant's own countermeasures (as we introduced in Chap. 3), if during the attacker's attack and intrusion procedure, the patroller is patrolling in the plant, then the attacker would evidently have a probability of being detected. After patrolling during a specified period of time in a plant, the patrolling team moves to another plant belonging to the geographical cluster, via the (public) road. However, the attacker may know the patroller's daily patrolling routes, for instance, by long-term observation or by stealing the patroller's security plan.

6.2.2   Formulating the Research Question

6.2.2.1   Graphic Modelling

A chemical cluster can be described as a graph $G(V, E)$ where V represents the number of vertices (or nodes) of the graph, and E is the number of edges of the graph. The vehicle entrances of every plant and the crossroads that are situated on the road form the nodes of the graph. The roads between different plants (to be more acurrate, it should be "between different entrances") are modelled as edges of the graph. Furthermore, all entrance nodes which belong to the same plant are modelled to be fully connected, which means edges also exist between every two nodes in these cases.

Fig. 6.1 Layout of a chemical park in Antwerp port

Fig. 6.2 Graphic modelling
of the chemical park

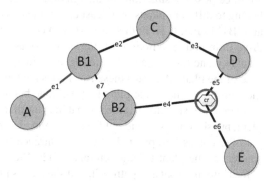

For example, Fig. 6.1 gives the layout of a small part of the Antwerp port chemical industrial park. There are five plants in this picture, indexed as plant 'A', plant 'B', and so forth. The yellow dot lines demonstrate the roads, which is the only infrastructure where the patroller can drive. Figure 6.2 shows the graph model of the cluster shown in Fig. 6.1. As we may notice, plants 'A', 'C', 'D', and 'E' in Fig. 6.1 are modelled as a node (with the same name) in Fig. 6.2. The cross point of the vehicle road between plant 'D' and 'E' in Fig. 6.1 is also denoted as a node in Fig. 6.2 (i.e., node 'cr'). Moreover, plant 'B' has two vehicle entrances, and therefore two nodes (i.e., nodes 'B1' and 'B2') are used in Fig. 6.2 to denote these two different entrances of plant 'B'. Edges 'e1' to 'e6' reflect the vehicle roads between different plants, while edge 'e7' is added between node 'B1' and 'B2' because these two nodes belong to the same plant and hence should be connected.

Based on the graphic model, the patrolling scenario in Sect. 6.2.1 can be described as a graphic patrolling problem: (1) a patroller (team) starts her patrolling from a node (the base camp); (2) she moves in the graph; (3) when arriving at a node, she may decide whether to stay at the node for a specific period of time t_k^p (i.e., patrol the plant) or not (i.e., move to another plant without patrolling the current plant); (4) after a period T, the patroller terminates the patrolling and goes back to her base camp.

In the above statement, t_i^p represents the patrolling time in plant i. t_i^p is determined both by the plant and by the patrolling scenario. For instance, territorially big plants may have a longer t_i^p. Moreover, the patroller, if coming into plant i, can also have several different patrolling intensities, and more intensive patrolling needs a longer t_i^p, and vice versa. t_i^p would also be slightly influenced by the entrances where the patroller comes into, and also leaves, the plant. In this chapter, for the sake of simplicity, we assume that each plant has a fixed t_i^p, without considering the influence of different entrances and without considering the multiple patrolling intensities. T represents the total patrolling time, and its typical value can be, for instance, 3 h. Table 6.1 further demonstrates all the notations used in this chapter.

A superior connection matrix sC of graph G is defined. The entry $sC(i, j)$ denotes the time needed for the patroller to move from node i to node j (of graph G). There are three possible situations of the relationship of nodes i and j: (i) these two nodes belong to different plants or at least one of them is a cross road node (e.g., nodes 'A' and 'B1' in Fig. 6.2). In this case, $sC(i, j)$ equals the time that the patroller needs to drive from node i to node j. (ii) these two nodes are different entrances of the same plant (e.g., nodes 'B1' and 'B2' in Fig. 6.2). In this case, $sC(i, j)$ equals the patrolling time of the plant. And (iii) these two nodes are the same. In this case, $sC(i, j)$ equals the patrolling time of the plant that the node belongs to.

In practice, situation (ii) means that the patroller comes into a plant and patrols the plant, but she comes in and out from different entrances. For instance, in Fig. 6.2, the patroller comes into plant 'B' through entrance 'B1' and after patrolling plant 'B', she leaves the plant through entrance 'B2'. Situation (iii) means that the patroller comes into the plant and patrols it, and she comes in and out using the same entrance/exit gate. For instance, in Fig. 6.2, the patroller comes into plant 'B' through entrance 'B1' and after patrolling the plant, she leaves the plant through entrance 'B1' again.

Ideally speaking, the patroller may also pass a plant without patrolling it, for a purpose of shortening the traveling time of arriving at her next patrolling plant. In the cluster shown in Figs. 6.1 and 6.2, if the patroller wants to move from plant 'A' to plant 'E', instead following the route "A→B1→C→D→cr→E", she may also go the route "A→B1→B2→cr→E" without patrolling plant 'B'. In the latter route, since plant 'B' is not patrolled, the time needed from entrance 'B1' to entrance 'B2' can be quite short, resulting in a shorter traveling time for the latter route than the former route. However in practice, this behaviour (e.g., passing the plant without patrolling it) increases the risk for the passing-by plant (e.g., plant 'B' in the above example), because of that more people come into the plant. Therefore, unless an agreement exists, the patroller would not be allowed to pass a plant without patrolling it. Therefore, situation (ii) in this research is assumed to only represent the case that the patroller patrols the plant.

Table 6.1 Definitions of notations

Notation	Definition	Type[a]
$G(V, E)$	The graphic model of the chemical cluster, defined in Sect. 6.2.2.1.	MG
t_e^d	The patroller's travelling time on edge $e \in E$.	IN
t_i^p	The patroller's patrolling time within plant i.	IN
k^i	The intrusion and attack continuing time, in plant i.	IN
T	Total patrolling time.	IN
$pG(pV, pE)$	The patrolling graph of the chemical cluster, defined in Sect. 6.2.2.2.	MG
$|V|$	Nodes number of graph G.	MG
sC	Superior connection matrix of graph G.	MG
$dis(bcn, nd)$	The shortest distance (in time) in the graph G from the base camp node bcn to node $nd \in G$.	MG
c_{s-e}	Probability that the patroller takes the action represented by edge $(s, e) \in pE$.	MG
R^d	The patroller's reward by catching an attacker.	IN
L^d	The patroller's loss if an attack is succeed.	IN
P^a	The attacker's penalty if being caught.	IN
G^a	The attacker's gain from a successful attack.	IN
$f_{cpp}, \tilde{f}_{cpp}$	Probability that the attacker would be detected by the countermeasures of the plant, estimated from the defender and from the attacker's perspective respectively.	IN
f_p	Probability that he attacker would be detected by the patroller.	MG
f	Probability that the attacker would be detected.	MG
σ_r	Probability that the patroller would detect the attacker in overlap situation r, defined in Sect. 6.3.4.	IN
τ_r	Probability that he patroller would be in the overlap situation r, defined in Sect. 6.3.4.	MG
$s_a (s_d)$	An attacker (defender) pure strategy.	MG
$S_a (S_d)$	Strategy set of the attacker (defender).	MG
$\vec{c}$	The vector form of representing a defender's strategy.	MG
sP_{pv}	The probability that the patroller would be at node $pv \in pV$.	MG
cP_{pv}^{pe}	The conditional probability that patroller would take the action $pe \in pE$, in condition that she currently locates at $pv \in pV$.	MG

[a]IN means model inputs, and this kind of data should be provided by security experts; MG means model generated data

For the cluster and the graph shown in Figs. 6.1 and 6.2, if we set: $t_1^d = 2, t_2^d = 3, t_3^d = 4, t_4^d = 3, t_5^d = 2, t_6^d = 2$, and further set $t^p('A', 'B', 'C', 'D', 'E') = [9, 7, 6, 5, 7]$, then the superior matrix sG of the example can be shown in Table 6.2. t_i^d represents the driving time of edge 'ei' in Fig. 6.2. For instance, t_1^d is the driving time from node 'A' to 'B1'. $t^p('X')$ denotes the time needed to patrol plant 'X'. All the time-related data are unified in minutes.

Table 6.2 Superior connection matrix for Fig. 6.2 with the illustrative numbers

	A	B1	B2	cr	C	D	E
A	9	2	∞	∞	∞	∞	∞
B1	2	7	7	∞	3	∞	∞
B2	∞	7	7	3	∞	∞	∞
cr	∞	∞	3	∞	∞	2	2
C	∞	3	∞	∞	6	4	∞
D	∞	∞	∞	2	4	5	∞
E	∞	∞	∞	2	∞	∞	7

Table 6.3 An algorithm of generating the patrolling graph

Algorithm: Generating the patrolling graph
1. *Construct an empty temporary node list tNL, an empty node list pV, an empty edge set pE;*
2. *Construct node pv = (0, bcn), in which bcn is the patrolling base camp node in graph G;*
3. *Initialize tNL ← pv, pV ← pv;*
4. *While tNL not empty, do*
4.1. *Get the first node in tNL, denoted as the current node cv = (ct, cn);*
4.2. *Construct follow-up nodes of cv;*
4.2.1. *In graph G, find all the connected nodes of cn, representing as ccn = {nd ∈ V∣sC (cn, nd) < ∞};*
4.2.2. *For each nd ∈ ccn, if ct + sC(cn, nd) ≤ T + dis(bcn, nd), construct a new node nv = (ct + sC(cn, nd), nd) and a directed edge ne from cv to nv should also be constructed;*
4.2.3. *Add ne to pE;*
4.2.4. *if nv in pV already, continue; otherwise, insert nv into tNL, add nv to pV;*[a]
4.3. *Remove cv from tNL*
5. *End*

[a]*tNL should be sorted according to the nodes' time*

6.2.2.2 Patrolling Graph Modelling

A directed patrolling graph $pG(pV, pE)$ is defined based on the graphic model of the chemical cluster. A node of pG is defined as a tuple of (t, i), in which $t \in [0, T]$ denotes the time dimension and $i \in \{1, 2, \ldots, |V|\}$ denotes a node in graph $G(V, E)$ (i.e., a plant (entrance) in the chemical cluster). Node (t, i) means that at time t the patroller arrives or leaves node i. A directed edge of pG from node (t_1, i_1) to node (t_2, i_2) therefore denotes a patroller action where she moves from node i_1 at time t_1 to node i_2, and arrives at t_2. Table 6.3 shows an iterative algorithm for generating the patrolling graph $pG(pV, pE)$. $dis(bcn, nd \in G)$ is the shortest distance (in time) in graph G from the base camp node bcn to node nd.

Figure 6.3 shows the patrolling graph pG for the chemical cluster shown in Fig. 6.1, with the data in Table 6.2 and further assuming a patrolling time $T = 30$. The patroller's base camp is assumed to be close to the cross road node, thus 'cr' is chosen as the patroller's base camp.

In Fig. 6.3, the x axis denotes the time dimension, while the y axis represents the different nodes in Fig. 6.2. Therefore, any coordinates in Fig. 6.3 can be a possible

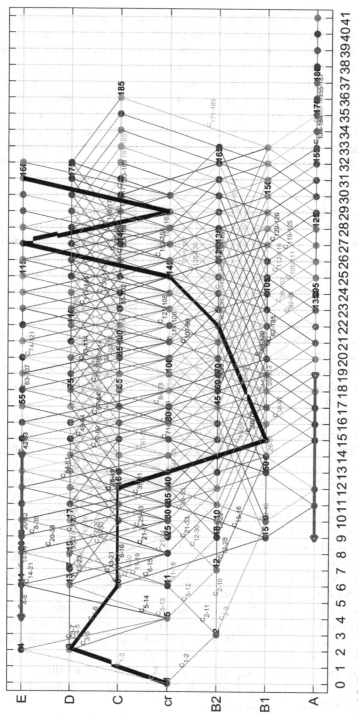

Fig. 6.3 Patrolling Graph of the illustrative example

node for pG. As we may see, node 1 (at the left-hand side of the figure) in Fig. 6.3 is $(0,\,'cr')$, which means that at time 0, the patroller starts from her base camp (i.e., 'cr'). Thereafter she has 3 choices: (i) to come to plant 'B' (more accurately, entrance 'B2') with a driving time t_4^d, and reaches node 2 (i.e., (3,'B2')); (ii) to come to plant 'D' with a driving time t_4^d, and reaches node 3 (i.e., (2,'D')); and (iii) to come to plant 'E' with a driving time t_6^d, and reaches node 4 (i.e., (2,'E')). Subsequently, at new nodes (e.g., 2, 3, or 4), the patroller has the same choice problem, that is, to patrol the current plant or to come to another plant. Finally, when time comes to the end of the patrol, the patroller terminates the patrol and comes back to her base camp. In Fig. 6.3, the indexes of some nodes and the weight of some edges are not shown, for the purpose of improving the visibility of the figure. Furthermore, the actions (edges) that the patroller comes back to her base camp are not shown, since these actions do not have an influence on the patrolling results.

A fixed patrolling route is a series of edges $(pe^1, pe^2, \ldots, pe^{len})$ in the patrolling graph that satisfies the following three conditions: (i) the in-degree of the start node of pe^1 is 0; (ii) the out-degree of the end node of pe^{len} is 0; and (iii) pe^i and pe^{i+1} $(i = 1, 2, \ldots, len - 1)$ are linked, which means that the end node of pe^i is the start node of pe^{i+1}. For instance, the bold (and black) line in Fig. 6.3 denotes a fixed patrolling route, and it is: $'cr' \rightarrow 'D' \rightarrow 'C' \rightarrow$ patrol plant $'C' \rightarrow 'B1' \rightarrow$ patrol plant $'B' \rightarrow$ leave plant $'B'$ from $'B2' \rightarrow 'cr' \rightarrow 'E' \rightarrow 'cr' \rightarrow 'E'$.

A purely randomized patrolling route is defined as: "at a node of the patrolling graph, the patroller goes to each edge outgoing from the node with an equal probability." For instance, in Fig. 6.3, at node 1 $(0,\,'cr')$, the patroller goes to node 2, 3, or 4 with a probability 1/3, and at node 2 $(3,\,'B2')$, the patroller goes to node 9, 10, or 11 all with a probability 1/9, and so forth.

To keep the continuity of coverage of each plant, the patroller is required to prolong her patrolling in the plant until the next patroller team might be able to arrive at the plant (see step 4.2.2 in Table 6.3). For instance, in Fig. 6.3, though the patrolling time is set as $T = 30$, however, the patrolling in plant 'A' is not stopped until $t = 41$. The idea is that, the shortest time that the next patrolling team can arrive at plant 'A' (from 'cr') is 11 (By following a path $'cr' \rightarrow 'B2' \rightarrow 'B1' \rightarrow 'A'$). If the current patroller team does not prolong her patrolling, and the next patroller team starts at time 30 and starts from her base camp (i.e., 'cr'), then plant 'A' would definitely not be covered during time $(30, 41)$. This approach may increase the patroller's workload. However, if we set T slightly smaller than the patroller's real workload, the problem will be solved. For example, if a patroller team's workload is 240 min per day, for modelling reasons we set it at $T = 220$.

The way that we deal with the continuity of patrolling coverage (or the periodic patrolling problem) implies that during time $[T, T + \max\,(dis(bcn, nd \in G))]$, there might be two patrolling teams in the industrial park at the same time. Nevertheless, in each plant, there is maximally one patrolling team present. The second patrolling team starts from her base camp at time T and also probabilistically schedules her actions according to the patrolling graph. Therefore, the time period of $[30, 41]$ of Fig. 6.3 should actually be an overlap.

6.2.2.3   Time Discretization

The time dimension (x axis) of the patrolling graph is continuous. Therefore, the patroller's traveling time t^d and patrolling time t^p are not necessarily integers. Moreover, the adversary's attack can happen at any time belonging to the continuous time interval $[0, T)$.

In our model, we discretize the time dimension of the patrolling graph. The time interval $[0, T)$ is divided to be multiple equal time slices and the length of each time slice can be, for instance, a second or a minute. All the time-related parameters (e.g., the patroller's traveling time and patrolling time, the attacker's attack period) are rounded to their closest integer numbers of the time slice. For instance, if there is a t^p $= 6.3$ *minutes* and the time slice is defined as 1 *minute*, then we would have $t^p = 6$. Moreover, the attacker can only start his attack at the beginning of each time slice and his attack period lasts for several time slices. Consequently, any action of the patroller and the attacker would happen at the beginning points of each time slice, and we therefore denote the time interval $[0, T)$ as $\{0, \ldots, \bar{T} - 1\}$, of which the latter means all the non-negative integers smaller than $\bar{T}$, and $\bar{T}$ is the number of time slices.

Discretization of the time axis simplifies the model. As we will see in Sect. 6.3.2, by discretizing the time axis, all the attacker's actions can be enumerated. Furthermore, discretizing the time axis also makes it easier to calculate the detection probabilities, as shown in Sect. 6.3.4. Discretization of the time dimension is also reasonable from a practical point of view. Although time is continuous in reality, we would stop at a certain accuracy, for instance, at seconds. Therefore, if the length of a time slice is short enough, the discretization model describes the reality very well.

6.3   Game Theoretic Modelling

6.3.1   Players

Players of the chemical cluster patrolling (CCP) game are the patroller team on the one hand (defender) and the potential adversaries on the other (attacker). The CCP game is a two players game and both players are assumed to be perfectly rational.

6.3.2   Strategies

Attacker Strategy
An attacker's strategy consists of three parts: (i) determine a target plant to attack; (ii) determine a time to start the attack; and (iii) determine an attack scenario to use. Different attack scenarios may need different intrusion and attack efforts, resulting in

different attack continuing times. For instance, generally speaking, an attack scenario with a suicide bomber needs less time than an attack scenario aiming to steal hazardous materials from the chemical plant, since there is an exit step for the latter scenario.

An attacker's pure strategy can be denoted as Formula (6.1).

$$s_a = (t, i, k_i) \tag{6.1}$$

In which t denotes the attack start time, i represents the target plant, k_i is the attack continuing time (e.g., 10 min) which should be determined by both the attack scenario and the target plant.

Example: the two horizontal bold dot red lines in Fig. 6.3 represent attacks of attacking plant 'A' at time 9 (the horizontal line at the bottom) and of attacking plant 'E'at time 4 (the horizontal line at the top), with an intrusion and attack continuing time of ten time units, respectively.

Formula (6.1) implies that the attacker would only attack one plant. The number of the attacker's pure strategies can be calculated by Formula (6.2). In which m is the number of pure strategies of the attacker; n denotes the number of plants in the cluster; T is the total number of time slices; and Sce is the number of different attack scenarios.

$$m = n \cdot T \cdot Sce \tag{6.2}$$

Patroller Strategy

The patroller's strategy is to randomize her patrolling and to bring maximal uncertainties (about her location) to the attacker. According to the patrolling graph we constructed in Sect. 6.2.2, at each node of pG, the defender may choose to patrol the current plant or move to other adjacent plants, and her choices are represented as the edges in pG. Therefore, if we assign a probabilistic number to each edge of pG, and define the number as the probability that the defender may go to that edge (please recall the meaning of an edge in pG, as stated in Sect. 6.2.2), then the patroller's strategy is the combination of these probabilistic numbers. A mathematic formulation of the defender's strategy is shown in Formula (6.3).

$$s_d = \prod_{(s,e) \in pE} c_{s-e} \tag{6.3}$$

In which c_{s-e} denotes the probabilistic number assigned to the edge from node s to node e, $\prod$ denotes the Cartesian product of all edges in pG (i.e., all $(s, e) \in pE$).

An intermediate node of pG is a node that has both income edges and outcome edges. A root node of pG is a node that has no income edges. For instance, node $(0, cr)$ in Fig. 6.3 is a root node, but not an intermediate node, while node $(2, D)$ is an intermediate node but not a root node. An important property of probabilities c_{s-e} is that, for each intermediate node (of pG), the sum of all the income probabilities must equal the sum of all the outcome probabilities. This is a result of the definition of the probabilities. The sum of all the income probabilities (of a node) represents how likely the patroller will be at the node, while the sum of all the outcome probabilities represents the probability that the patroller would take an action (either goes to adjacent plants or patrols the current plant) at the node. Another property of probabilities c_{s-e} is that, the sum of probabilities coming out from the root node equals 1. The idea behind this property is that, the patroller deterministically (since a probability of 1) starts from the root node, and then she chooses to go to the next step. Formulas (6.4) and (6.5) illustrate the abovementioned two properties.

$$sP_{pv} = \sum_{in\in\{s\in pV|(s,pv)\in pE\}} c_{in-pv} = \sum_{out\in\{e\in pV|(pv,e)\in pE\}} c_{pv-out} \qquad (6.4)$$

$$\sum_{out\in\{e\in pV|(root,e)\in pE\}} c_{root-out} = 1 \qquad (6.5)$$

Furthermore, in patrolling practice, when the defender is already situated at node pv (of pG), her conditional probability of choosing a specific action (i.e., an edge in pG) can be calculated by Formula (6.6). For instance, if a purely randomized patrolling strategy would be implemented on the patrolling graph shown in Fig. 6.3, then the probability that the patroller will be at node 2 (3, 'B2') is $sP_2 = 1/3$, and the probabilities that the patroller goes to node 9, 10, and 11 are all $c_{2-9} = c_{2-10} = c_{2-11} = 1/9$. Therefore, we have $cP_2^9 = cP_2^{10} = cP_2^{11} = 1/3$, and this result means that at node 2, the patroller takes each action at the same probability. Fig 7.12 in Chap. 7 also illustrates how Formula (6.6) works.

$$cP_{pv}^{out} = \frac{c_{pv-out}}{sP_{pv}}, \text{ for all } out \in \{v\in pV|(pv,v)\in pE\} \qquad (6.6)$$

6.3.3 Payoffs

There are two possible results in the CCP game, being: (i) the attack fails, either stopped by the multiple-plant patroller team or by the countermeasures in the target plant and (ii) the attack is successfully implemented. If the attack fails, the patroller gets a reward R^d (e.g., obtaining a bonus) and the attacker suffers a penalty P^a (e.g., being sent to prison). If the attacker succeeds, the patroller suffers a loss L^d and the attacker obtains a gain G^a.

R^d is a number decided by the chemical cluster council. For instance, the cluster rewards $1k€$ to the defender (consists of the patroller and the plant's own security department). P^a is scenario-related since different attack scenarios need different attack costs and the attacker, if being caught, will also be punished differently. L^d and G^a are determined by both the attack scenario and the target plant. All these parameters should be evaluated by security experts, for instance, by an API SRA team [7]

Formulas (6.7) and (6.8) further define the patroller and the attacker's payoff, in which $f(\tilde{f})$ is the probability that the attack would be detected, from the defender's (the attacker's) perspective.

$$u_d = R^d \cdot f - L^d \cdot (1 - f) \tag{6.7}$$
$$u_a = G^a \cdot (1 - \tilde{f}) - P^a \cdot \tilde{f} \tag{6.8}$$

6.3.4   Computing the Probability of the Attack Being Detected (f)

In this section, we focus on calculating the probability $f(\tilde{f})$ that the attacker would be detected, under the condition that the attacker plays a strategy (t, i, k_i) and the defender plays $\vec{c}$ (a vector whose entries are the c_{i-j} in Formula (6.3)).

We denote the probability that the security countermeasures in the target plant would detect the attacker as f_{cpp}, which can be calculated by the Chemical Plant Protection game proposed by Zhang and Reniers [Zhang, 2016 #953] or which can be evaluated by a security assessment team as well. Furthermore, we represent the probability that the patroller would detect the attacker as f_p. Considering that the attacker can be detected either by the countermeasures of the target plant or by the patroller team, the probability that the attacker would be detected can be calculated by Formula (6.9):

$$f = 1 - (1 - f_{cpp}) \cdot (1 - f_p) \tag{6.9}$$

Note that f_{cpp} is a plant-specific parameter (a real number belonging to [0,1]). We focus on calculating f_p. An intrusion and attack procedure in plant i lasts for k_i time slices, while patrolling in the plant lasts for t_i^p time units. If there are any overlaps between the intrusion and attack procedure and the patroller's staying in the plant, then there is a probability that the attacker would be detected by the patroller. Otherwise, the adversary would only possibly be detected by the countermeasures of the target plant, i.e., $f_p = 0$. Theoretically speaking, the longer the overlap is, the higher the f_p would be.

Furthermore, which time period of the intrusion and attack procedure is covered by the overlap also influences the probability. For instance, the intruder can easier be noticed by the patroller team at the beginning of his intrusion procedure since at this time, he is moving into the plant. After reaching the target, it may be difficult for a patroller to detect the attacker. For instance, if his target is inside a room, then the patroller would not be able to detect him at all. The situation can also be opposite.

Therefore, in order to calculate f_p, not only the length of the overlap should be calculated, but also which part of the intrusion and attack procedure is covered should also be identified. The overlap of the patroller's staying in plant i and the attacker's intrusion and attack procedure in plant i can be calculated by Formula (6.10), in which st denotes the start time that the patroller stays in plant i. There are two situations of the exact overlap period.

$$Overlap = \left[max\{t, st\}, min\{t + k_i, st + t_i^P\}\right] \qquad (6.10)$$

Situation 1: if $t_i^P \leq k_i$, then there are $k_i + t_i^P - 1$ possible overlap situations. Each situation covers the intrusion and attack procedure at time $[t, t + 1]$, $[t, t + 2]$, ..., $[t, t + t_i^P]$, $[t + 1, t + t_i^P + 1]$, ..., $[t + k_i - t_i^P, t + k_i]$, $[t + k_i - t_i^P + 1, t + k_i]$, $[t + k_i - t_i^P + 2, t + k_i]$, ..., $[t + k_i - 1, t + k_i]$, respectively. Figure 6.4 shows an example of the overlap situations with $k_i = 5, t_i^P = 2$. In Fig. 6.4, the horizontal line denotes the intrusion and attack procedure which lasts for 5 time slices, while the red dotted line indicates the overlap of the intrusion with the patroller's staying in the plant.

Situation 2: if $t_i^P > k_i$, then there are $k_i + k_i - 1$ possible overlap situations. Each of the situations cover the intrusion and attack procedure at time $[t, t + 1]$, $[t, t + 2]$, ..., $[t, t + k_i]$, $[t + 1, t + k_i]$, ..., $[t + k_i - 1, t + k_i]$, respectively.

For each of the possible overlap cases, define a detection probability σ_r, and $r = 1, 2, ..., t_i^P + k_i - 1$ in situation 1 and $r = 1, 2, ..., k_i + k_i - 1$ in situation 2. Furthermore, denote the probability that the patroller would be in situation r as τ_r. The probability that the attacker would be detected by the patroller can then be calculated by Formula (6.11). σ_r are user inputs and should be provided by security experts.

$$f_p = \sum_r \sigma_r \cdot \tau_r. \qquad (6.11)$$

Table 6.4 shows how to calculate τ_r, under the condition of an attacker strategy (t, i, k_i) and a defender strategy $\vec{c}$.

Denote the start and end node of an edge (of pG) as $sn = (snt, sni)$ and $en = (ent, eni)$ respectively, and define:

Condition 1: both the corresponding entrances of node sni and eni belong to plant i, the attacker's target. For instance, in the illustrative example shown in Figs. 6.1 and 6.2, if the target plant is 'A', and $sni = eni = \dot{A}$, then condition 1 holds; or if the target plant is 'B' and $sni = \dot{B1}$ and $eni = \dot{B2}$, then condition 1 holds as well.

Fig. 6.4 An illustrative figure of the overlap situation

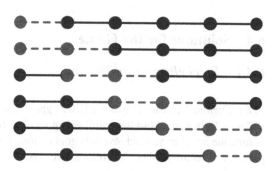

Table 6.4 The procedure of calculating τ_r

Calculating τ_r
1. Initialize $\tau_r = 0$.
2. If an edge $pe \in pE$ in the patrolling graph pG satisfies condition 1 and condition 2, then $\tau_r = \tau_r + c_{pe}$, in which c_{pe} is the weight (the probability) of the edge.

Condition 2: the overlap (in time dimension) of the edge and the attacker strategy satisfies situation r. Rigorously, $[snt, ent] \cap [t, t + k_i]$ equals the corresponding time zone of the overlap situation r. Fig 7.12 and Table 7.35 in Chap. 7 illustrate this condition.

In condition 1, if $sni = eni$, the edge would be a horizontal line when shown in a figure like Fig. 6.3, and it indicates that a patrolling team comes in and out the same gate of the plant, otherwise if $sni \neq eni$ but both of them belong to the same plant, it means that the patrolling team enters and leaves via different gates of the plant.

In condition 2, if $t + k_i > T$, then edges satisfying the condition that $[snt, ent] \cap [0, t + k_i - T]$ equals the corresponding time zone of the overlap situation r, are also said to fulfil condition 2. This results from the way that we deal with the periodic patrolling problem. When time exceeds T, the next patrolling team has already started her patrolling, therefore the attacker not only can be detected by the current patroller, but also can be detected by the next patrolling team.

It is worth noting that τ_r is a linear polynomial of $\vec{c}$, denoted as $\tau_r = Coe_r \cdot \vec{c}^T$, and f_{cpp} and σ_r are user provided parameters. Therefore, f is a linear polynomial of $\vec{c}$ as well. Furthermore, the definitions of f, u_d, u_a can be rewritten as:

$$f = \left[\sum_r (1 - f_{cpp}) \cdot \sigma_r \cdot Coe_r, f_{cpp} \right] \cdot \left[\vec{c}, 1 \right]^T \qquad (6.12)$$

$$u_d = \left[(R^d + L^d) \cdot \left(\sum_r (1 - f_{cpp}) \cdot \sigma_r \cdot Coe_r \right), (R^d + L^d) \cdot f_{cpp} - L^d \right] \cdot \left[\vec{c}, 1 \right]^T \qquad (6.13)$$

$$u_a = \left[-(G^a + P^a) \cdot \left(\sum_r (1 - \tilde{f}_{cpp}) \cdot \sigma_r \cdot Coe_r \right), G^a - (G^a + P^a) \cdot \tilde{f}_{cpp} \right] \cdot \left[\vec{c}, 1 \right]^T \qquad (6.14)$$

6.4   Solutions for the Game

6.4.1   Stackelberg Equilibrium

In the Chemical Cluster Patrolling (CCP) game, the attacker is assumed to be able to collect information about the patroller's patrolling route. For instance, as mentioned before, the attacker may achieve this goal by long term observation or by stealing the patroller's security plan. Therefore, we assume that the CCP game is played sequentially. The patroller (being the game leader) firstly commits a patrolling strategy $\vec{c}$, and subsequently, the attacker moves optimally according to the defender's strategy (being the game follower). The patroller could also work out the attacker's optimal solution, and so she can arrange her strategy $\vec{c}$ optimally as well.

A Stackelberg equilibrium $(s_d^*, s_a^*) = \left(c^*, (t^*, i^*, k_i^*)\right)$ for the CCP game is a patroller-attacker strategy pair that satisfies the following condition:

$$(t^*, i^*, k_i^*) = \operatorname{argmax}_{(t,i,k_i)\in S_a}\left\{u_a\left(\vec{c}, (t, i, k_i)\right)\right\} \tag{6.15}$$

$$c^* = \operatorname{argmax}_{\vec{c}\in S_d}\left\{u_d\left(\vec{c}, (t^*, i^*, k_i^*)\right)\right\} \tag{6.16}$$

Formula (6.15) indicates that observing the defender's strategy $\vec{c}$, the attacker would play a strategy which maximizes his own payoff (i.e., a best response strategy). Formula (6.16) represents that the defender can also work out the attacker's best response to her strategy, thus she plays accordingly.

By discretizing the time dimension (in Sect. 6.2.2.3), the attacker has a finite number of strategies. Moreover, Formulas (6.13) and (6.14) show that for a given attacker strategy, payoff functions u_a and u_d would both be linear polynomials of $\vec{c}$. Therefore, a multiple linear programming algorithm[8] can be introduced to compute the Stackelberg equilibrium for the CCP game, as shown in Table 6.5.

In the linear programming step, the defender is solving a linear programming problem. The cost function of the linear programming is given by Formula (6.18) and the constraints are given by Formulas (6.17), (6.4) and (6.5). Furthermore, the LP step should be implemented for each attacker strategy. In the linear programming step, if we further constraint c_{s-e} to be either 0 or 1, then the MultiLPs algorithm would output the optimal fixed patrolling route for the patroller.

Table 6.5 MultiLPs algorithm for computing the Stackelberg equilibrium for the CCP game

MultiLPs
Initialization
For each attacker strategy (t, i, k_i), calculate u_a and u_d, which are linear polynomials of $\vec{c}$;
Linear Programming (LP)
Suppose that the attacker strategy $(t^\#, i^\#, k_i^\#)$ is the attacker's best response, which means:
$u_a\left(t^\#, i^\#, k_i^\#, \vec{c}\right) \geq u_a\left(t, i, k_i, \vec{c}\right), \forall (t, i, k_i) \in S_a$ (6.17)
The defender would then aims at:
$Pof_d\left(t^\#, i^\#, k_i^\#, \vec{c}^\#\right) = \max\limits_{\vec{c}\in S_d} u_d\left(t^\#, i^\#, k_i^\#, \vec{c}\right)$ (6.18)
Summary
The Stackelberg equilibrium $\left(c^*, (t^*, i^*, k_i^*)\right) = \arg\max\limits_{(t^\#,i^\#,k_i^\#)\in S_a} Pof_d\left(t^\#, i^\#, k_i^\#, \vec{c}^\#\right).$

The Stackelberg equilibrium calculated by the MultiLPs algorithm is a Strong Stackelberg Equilibrium, and it is therefore based on the "breaking-tie" assumption, as illustrated in Sect. 2.2.3. By running again the LP step in the MultiLPs algorithm, and supposing that (t^*, i^*, k_i^*) is the attacker's best response as well as revising Formula (6.17) to be Formula (6.19), in which α is a constant small positive number, the Strong Stackelberg Equilibrium will be slightly modified, resulting in a Modified Stackelberg Equilibrium which does not rely on the "breaking-tie" assumption and is still optimal enough.

$$u_a\left(t^*, i^*, k_i^*, \overrightarrow{c}\right) \geq \alpha + u_a\left(t, i, k_i, \overrightarrow{c}\right), \forall (t, i, k_i) \in S_a \qquad (6.19)$$

6.4.2   Robust Solution Considering Distribution-Free Uncertainties

The Stackelberg Equilibrium can be calculated for the CCP game only in case that the patroller knows the exact numbers of all the parameters (shown in Table 6.1) of the game. In security practice, the patroller may obtain some of these parameters by using conventional security risk assessment methods such as the API SRA [7] However, there are at least two parameters of which the values are difficult to obtain: the attacker's gain from a successful attack G^a and the attacker's estimation of being detected by the intrusion detection system of each plant $\tilde{f}_{cpp}$. Therefore, similar to the Interval CPP Game we defined in Chap. 4, we assume that the patroller can obtain an interval of these two parameters and how these two parameters distribute in the interval zones are not known. Further assume that $G^a \in [G^{a\_min}, G^{a\_max}]$ and $\tilde{f}_{cpp} \in \left[\tilde{f}_{cpp}^{min}, \tilde{f}_{cpp}^{max}\right]$ and therefore the patroller can have the lower and upper bound of the attacker's payoff, as shown in Formulas (6.20) and (6.21) respectively. Note that Formula (6.8) demonstrates that u_a is monotonically increasing on G^a and monotonically decreasing on f. Formula (6.9) demonstrates that f is monotonically increasing on f_{cpp}.

$$u_a^{min}$$
$$= \left[-\left(G^{a\_min} + P^a\right) \cdot \left(\sum_r \left(1 - \tilde{f}_{cpp}^{max}\right) \cdot \sigma_r \cdot Coe_r\right), G^{a\_min} - \left(G^{a\_min} + P^a\right) \cdot \tilde{f}_{cpp}^{max}\right]$$
$$\times \cdot \left[\overrightarrow{c}, 1\right]^T$$

$$(6.20)$$

$$u_a^{max}$$
$$= \left[-(G^{a\_max} + P^a) \cdot \left(\sum_r \left(1 - \tilde{f}_{cpp}^{\ min}\right) \cdot \sigma_r \cdot Coe_r\right), G^{a\_max} - (G^{a\_max} + P^a) \cdot \tilde{f}_{cpp}^{\ min}\right]$$
$$\times \cdot \left[\vec{c}, 1\right]^T$$

$$(6.21)$$

Knowing the lower and upper bound of the attacker's payoff, the patroller can play the game as follows: (i) she commits to a patrolling strategy c; (ii) she works out the attacker's lower and upper bound payoffs in the case that the attacker responds with different pure strategies to c; (iii) she gets the attacker's highest lower bound payoff R; (iv) she picks out all the attacker's possible best responses, which are, the attacker's pure strategies that have higher upper bound payoffs than R; (v) among all the attacker's possible best responses, assume that the one that is worst to the patroller is the attacker's real best response and the patroller then optimizes c accordingly. This procedure is the same as we used in the Interval Chemical Plant Protection Games in Sect. 4.3.

Furthermore, if two pure strategies of the attacker (e.g., s_{a1} and s_{a2}) have the same target plant, then the attacker's payoffs by responding these two pure strategies will share the same G^a and $\tilde{f}_{cpp}$ and therefore the payoffs (of responding these two strategies) will be correlated. In this situation, we have that $u_a(s_{a1}, c) \geq u_a(s_{a2}, c) \Leftrightarrow f_p(s_{a1}, c) \leq f_p(s_{a2}, c)$ and vice versa.

Formula (6.22) illustrates an algorithm for calculating the patroller's robust solution considering her distribution-free uncertainties on the attacker's parameters. In Formula (6.22), the variables are c, q, R and γ, which denote the patroller's patrolling strategy, indication of the attacker's possible best response strategy, the attacker's highest lower bound payoff, and the defender's optimal payoff, respectively.

$$\begin{aligned}
&\underset{c,q,R,\gamma}{\text{maximize}\,\gamma} \\
&s.t. \begin{cases}
c1. & NstFlwLef \cdot c = NstFlwRgt \\
c2. & R = u_a^{min}(J, c) \\
c3. & R \geq u_a^{min}(j, c), \forall j \in \{1, 2, \ldots, m\} \\
c4. & -q_j \cdot \Gamma \leq R - u_a^{max}(j, c) \leq (1 - q_j) \cdot \Gamma, \forall j \in \{1, 2, \ldots, m\} - Plt_J \\
c5. & -q_j \cdot \Gamma \leq f_p(j, c) - f_p(J, c) \leq (1 - q_j) \cdot \Gamma, \forall j \in Plt_J \\
c6. & (1 - q_j) \cdot \Gamma + u_d(j, c) \geq \gamma, \forall j \in \{1, 2, \ldots, m\} \\
c7. & q_j \in \{0, 1\}, c_i \in [0, 1], \forall j \in \{1, 2, \ldots, m\}, \forall i \in \{1, 2, \ldots, n\}
\end{cases}
\end{aligned}$$

$$(6.22)$$

Constraint $c1$ reflects the features of the patroller's strategy c, as explained in Formula (6.4) and (6.5). Constraints $c2$ calculates the attacker's lower bound payoff R by playing strategy J. $c3$ ensures that strategy J has the highest lower bound payoff, among all the attacker's pure strategies. Constraints $c4$ and $c5$ pick out all the attacker's possible best responses. Plt_J denotes all the attacker's strategies that have the same target plant with strategy J. Note that in these two constraints, if $u_a^{max}(j, c) > R$ or $f_p(j, c) < f_p(J, c)$, then $q_j = 1$, and vice versa. Therefore $q_j = 1$ indicates that strategy j is in the attacker's possible best response set. Constraint $c6$ represents the patroller conservatively thinking that among all the attacker's possible best responses, the one that is the worst to her is the attacker's real best response. The cost function further represents the patroller optimizing her payoff.

In Formula (6.22), the attacker's strategy J is assumed to have the highest lower bound payoff. Therefore, the optimal solution and payoffs generated by the formula are conditional. By implementing Formula (6.22) for m times and each time setting a different J, we obtain a result, denoted as $rlt^J = (c^J, q^J, R^J, \gamma^J)$. If Formula (6.22) is not feasible for a certain J, then we set $rlt^J = (null, null, -inf, -inf)$. Finally, we pick out the rlt^J that has a highest γ^J as the final robust solution of the game.

6.4.3   Robust Solutions Considering Implementation Errors and Observation Errors

Besides the defender's uncertainties on the attacker's parameters, there are other two types of uncertainties, namely, the patroller's implementation error and the attacker's observation error.

In reality, the patroller would always have errors while implementing her patrolling strategy. For instance, the patroller may have to go to the toilet or she has to deal with some detected security issues. Therefore, to make the patrolling strategies generated by the CCP game more robust, we can assume that the real patrolling strategy c^{real} may deviate slightly from the planned strategy c, that is, $c^{real} \in [c - \epsilon, c + \epsilon] \cap [0, 1]$, in which ϵ is a small positive number denoting the tolerance of the implementation error.

The attacker's observation error of the patroller's implemented strategy can be modelled in two different approaches. The first approach is similar to the modelling of the patroller's implementation error by introducing a small positive number δ, denoting the error between the attacker's observation and the defender's implemented strategy. Subsequently, we have $c^{obs} \in [c^{real} - \delta, c^{real} + \delta] \cap [0, 1]$. The second approach is by employing the anchoring theory. The anchoring theory says that when there is no external information about a set of discrete events, humans assume that the occurrence probability of each event is the same. When further information is provided (e.g., the attacker observes the patroller's daily patrolling), humans are able to calibrate their estimation of probability of each event to the real probability. In the CCP game, this procedure can be described as $c^{obs} = (1 - \beta) \cdot c^{PureRandom} + \beta \cdot c^{real}$, in which β denotes the observation ability of the attacker and $c^{PureRandom}$ denotes a purely randomized patrolling strategy.

For integrating these two types of uncertainties to the CCP game, the algorithm proposed by Nguyen et al.[9] can be employed. However, the algorithm in Nguyen et al.[9] has a very high computational complexity if being applied on the CCP game. Therefore, developing a quicker and more efficient algorithm for dealing with these two types of uncertainties in the CCP game can be a fruitful future research.

6.5 Conclusion

Terrorism is a global problem. Geographically clustered chemical plants throughout the world can be quite interesting targets for terrorists, due to the possibility of inducing domino effects. Besides the countermeasures that each plant takes, and that a multi-plant council may take, also security patrolling at the cluster level is recommended. To this end, a so-called chemical cluster patrolling game (CCP game) is developed and proposed in this chapter. The game is played by the patroller and the potential attackers, taking into account intelligent interactions between them. Two solution concepts, namely the Stackelberg Equilibrium and the 'robust solution', are put forward.

References

1. Reniers G, Pavlova Y. Using game theory to improve safety within chemical industrial parks. London: Springer; 2013.
2. Shieh E, An B, Yang R, Tambe M, Baldwin C, DiRenzo J, et al., editors. Protect: a deployed game theoretic system to protect the ports of the United States. In: Proceedings of the 11th international conference on autonomous agents and multiagent systems-volume 1. International Foundation for Autonomous Agents and Multiagent Systems; 2012.
3. Fang F, Stone P, Tambe M, editors. When security games go green: designing defender strategies to prevent poaching and illegal fishing. IJCAI; 2015.
4. Rezazadeh A, Zhang L, Reniers G, Khakzad N, Cozzani V. Optimal patrol scheduling of hazardous pipelines using game theory. Process Saf Environ Prot. 2017;109:242–56.
5. Alpern S, Morton A, Papadaki K. Patrolling games. Oper Res. 2011;59(5):1246–57.
6. Zhang L, Reniers G. CCP game: a game-theoretical model for improving chemical clusters patrolling. Accepted for publication in Reliability Engineering and System Safety; 2018.
7. API. Security risk assessment methodology for the petroleum and petrochemical industries. In: 780 ARP, editor. 2013.
8. Conitzer V, Sandholm T, editors. Computing the optimal strategy to commit to. In: Proceedings of the 7th ACM conference on Electronic commerce. ACM; 2006.
9. Nguyen TH, Jiang AX, Tambe M, editors. Stop the compartmentalization: unified robust algorithms for handling uncertainties in security games. In: Proceedings of the 2014 international conference on autonomous agents and multi-agent systems. International Foundation for Autonomous Agents and Multiagent Systems; 2014.

Chapter 7
Case Studies

Two case studies are carried out in this chapter, for illustrating the single plant protection game (the CPP game) and for demonstrating the multi-plant patrolling game (the CCP game), respectively.

7.1 Case Study #1: Applying the CPP Game to a Refinery

7.1.1 Case Study Setting

Figure 7.1 shows the layout of a refinery, which is also used as a case study in the API SRA document [1] and in Lee et al. [2].

The API SRA methodology concluded that the main gate (MG), the central control room (CCR), the co-gen unit and control room (CgCR), dock #1 (D1), and the tank farm (TF) are critical assets in this refinery, as described in the first page of form 1 in the API SRA document. We added the administration building (AdB), the electrical supply station (ESS), and the production facility (PF) to this list, since these assets also have important roles in the chemical plant defence in our opinion.

The main gate (MG) and Dock #1 (D1) are considered as critical assets since attackers may use these assets to intrude into the plant. Furthermore, D1 is located at the waterside and a tank farm (TF) is close to D1. If an attacker intrudes from D1 to attack the TF, the probability would be high and the consequence would be severe (especially the environmental damage). The MG is also vulnerable. Many people (the company's own employees as well as visitors of the plant) and vehicles are passing by the MG every day and an illegal intruder may try to act as one of the authorized people to intrude the plant. Countermeasures such as an access card system, CCTV, x-ray machine, and crash rated barrier etc. can therefore for instance be deployed at the MG and D1. The central control room (CCR) controls all production processes in the plant, and manages security communications in the

© Springer International Publishing AG, part of Springer Nature 2018   111
L. Zhang, G. Reniers, *Game Theory for Managing Security in Chemical Industrial Areas*, Advanced Sciences and Technologies for Security Applications,
https://doi.org/10.1007/978-3-319-92618-6_7

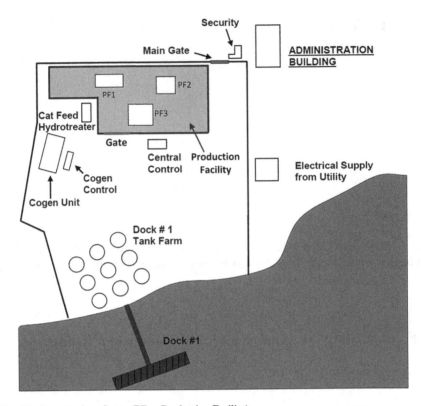

Fig. 7.1 Layout of a refinery (PF = Production Facility)

plant. If the CCR is damaged, there is a possibility for loss of lives, and a long recovering time is needed. The co-gen unit and control room (CgCR) provides steam for production and generates electrical power for the plant. There is limited redundancy in the electrical system. If the CgCR would be attacked, the plant would suffer at least a 2 days business interruption. The tank farm (TF) stores crude, intermediates, waste and finished liquid hydrocarbons. An attack on the TF may result in direct economic loss, environmental damage, casualties, business interruption, and what have you. We learned the above asset information from Form 1 of Example 2 in the API SRA document [1].

Furthermore, as already mentioned, we consider the production facility (PF), the administration building (AdB) and the electrical supply station (ESS) also as critical assets. If the production facilities would be attacked, besides the direct economic loss and possible casualties, the plant needs a long repairing time, resulting in business interruption for a long time. The administration building, although less vulnerable, may store important technique documents. Technique thieves can be quite interested in intruding into the building. The electrical supply station (ESS) provides electricity to the plant, and if being attacked, direct economic loss and business interruption would exist.

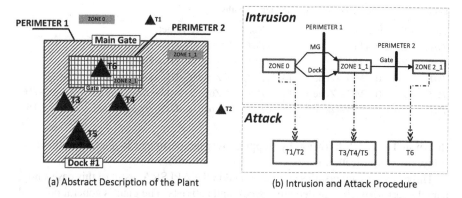

(a) Abstract Description of the Plant   (b) Intrusion and Attack Procedure

Fig. 7.2 Formalized representation of the refinery. (**a**) Abstract description of the plant (**b**) Intrusion and attack procedure

Table 7.1 Symbols map between Figs. 7.1 and 7.2a

Symbol in Fig. 7.2a	Symbol in Fig. 7.1
ZONE0	Outdoor area
ZONE1_1	Area within enclosure
ZONE2_1	Production facility area
PERIMETER 1	The boundary of the plant
PERIMETER 2	The boundary of the production facility
Main gate	Main gate
Dock #1	Dock #1
Gate	The entrance to the production facility
T1	Administration building
T2	Electrical supply from utility
T3	Cogen unit/Cogen control/Cat feed
T4	Central control
T5	Tank farm
T6	Production facilities in production facility area

Figure 7.2b shows the conceptual description of the refinery, based on the above analysis. Table 7.1 shows the corresponding names of symbols used in Fig. 7.1. Part of the perimeter 1 is the coastline, as we mentioned in Chap. 3 that a geographical border can also be a perimeter. There are multiple tanks in the tank farm and multiple production facilities (PF) in zone 2_1. However, on the one hand every tank or PF is similar in this illustrative example, and on the other hand, due to the possible existence of domino effects [3], if one tank or PF would be attacked, other tanks may also be damaged. Therefore, we simplify the whole tank farm and all the PFs as target 5 and target 6 respectively.

Figure 7.2b demonstrates the intrusion and attack procedure. For each attack scenario, there are in total ten possible combinations of intrusion paths and targets, as

Table 7.2 The attackers' intrusion and attack paths

No.	Path & Target	No.	Path & Target
PaT_1	Zone 0→T1	PaT_6	Zone 0→D1→Zone 1→T3
PaT_2	Zone 0→T2	PaT_7	Zone 0→D1→Zone 1→T4
PaT_3	Zone 0→MG→Zone 1→T3	PaT_8	Zone 0→D1→Zone 1→T5
PaT_4	Zone 0→MG→Zone 1→T4	PaT_9	Zone 0→MG→Zone 1→Gate→Zone 2→T6
PaT_5	Zone 0→MG→Zone 1→T5	PaT_{10}	Zone 0→D1→Zone 1→Gate→Zone 2→T6

shown in Table 7.2. Intrusion by stepping over the perimeter is ignored in this case study, for simplification reasons.

Three types of adversaries are mentioned in the API SRA result of this case study, namely, terrorists, disgruntled employees, and activists. The plant's general physical intrusion detection approach does not work for the disgruntled employees, thus this type of threat is excluded from this case study that we use further in this chapter. The terrorist mainly focuses on causing maximum damage to the plant as well as to the surrounding communities, while the activist is mostly concerned in shutting down the refinery operations. These two threats have threat levels 3 and 4 respectively (note that the threat level here only represents the likelihood (threat score) of the occurrence of the corresponding type of attacker, see also Fig. 1.6 in Chap. 1).

7.1.2 Chemical Plant Protection Game Modelling

7.1.2.1 Players

Players in this case study are the company security management team (e.g., an API SRA team), who acts as the defender, and the terrorist as well as the activist, who act as the possible adversaries. Therefore, the game would be a Bayesian game and the attacker has two different types. According to Formula (3.2), the prior probabilities that the attacker would be a terrorist and an activist are 3/7 and 4/7, respectively.

7.1.2.2 Strategies

A defender's pure strategy is the combination of different security alert levels (SAL) at each entrance and zone. In this case study, we assume that three different SALs (denoting as Green (1)/Yellow (2)/Red (3)) are possible at each entrance or zone. There are in total three entrances and three (sub-) zones in the case study, and we further list them as (i) Zone 0; (ii) the MG; (iii) D1; (iv) Zone 1; (v) the Gate; (vi) Zone 2. Subsequently, we employ a cross product of six digital numbers, i.e., $s_d = d_{Z0} \times d_{MG} \times d_{D1} \times d_{Z1} \times d_{Gate} \times d_{Z2}$, to denote a defender's pure strategy, of which d_{Z0} denotes the security alert level at Zone 0, d_{MG} denotes the security alert level at the Main Entrance, and so forth. For instance, an $s_d = 2 \times 1 \times 3 \times 2 \times 3 \times 1$

Table 7.3 The terrorist's pure strategy list

Index	Strategy
s_{v1}	$T1 \times VBIED$
s_{v2}	$T2 \times VBIED$
s_{v3}	$T3 \times MG \times VBIED$
s_{v4}	$T4 \times MG \times VBIED$
s_{v5}	$T5 \times MG \times VBIED$
s_{v6}	$T6 \times MG \times Gate \times VBIED$

Table 7.4 The activist's pure strategy list

Index	Strategy
s_{e1}	$T2 \times EA$
s_{e2}	$T3 \times MG \times EA$
s_{e3}	$T3 \times Dock \times EA$
s_{e4}	$T4 \times MG \times EA$
s_{e5}	$T4 \times Dock \times EA$
s_{e6}	$T6 \times MG \times Gate \times EA$
s_{e7}	$T6 \times Dock \times Gate \times EA$

represents that the SALs at Zone 0, the MG, D1, ZONE 1, the Gate, and Zone 2, are 2 (or Yellow), 1 (or Green), 3 (or Red), 2 (or Yellow), 3 (or Red), and 1 (or Green), respectively.

An attacker's pure strategy consists of a target, an intrusion path, and an attack scenario. The combinations of intrusion paths and targets are shown in Table 7.2. To simplify the case study, only one attack scenario is considered for each type of attacker. The terrorist is assumed to employ a vehicle-borne improvised explosive device (VBIED) as his attack scenario. We assume an environmental activist (EA) aiming to shut down the operation of the refinery as the scenario of the activist.

In a VBIED attack, a car or a truck should be involved. Therefore, the terrorist would not be able to intrude from Dock #1 (D1). The environmental activist aims to shut down the plant, instead of causing damages to the plant, thus we assume that the administration building and the tank farm would not be his attack target. Tables 7.3 and 7.4 illustrate all the pure strategies for the terrorist and for the activist respectively.

7.1.2.3 Payoffs

Tables 7.5 and 7.6 show the probabilities (i.e., P_i^z and P_r^p) that the attacker would successfully pass an entrance or a zone, for the terrorist and for the activist respectively. The "Typical" column shows the entrance or the zone. The "p_d" column shows the defender's estimation of the probability. The "$\tilde{p}_a$" columns (i.e., $\tilde{p}_a^{min}$, $\tilde{p}_a^{max}$, and $\tilde{p}_a^{nominal}$) illustrate the attacker's values of the probabilities. The data is given separately for the defender and the attacker, since they can evaluate the same parameter in different results, as explained in Chap. 3. In a sequential game setting,

Table 7.5 Basic probabilities of successful intrusion for the terrorist

Typical	From	To	p_d	$\tilde{p}_a^{min}$	$\tilde{p}_a^{max}$	$\tilde{p}_a^{nominal}$	Coe_2	Coe_3
zone0		T1 or T2	0.95	0.95	0.99	0.95	0.68	0.45
MG			0.3	0.3	0.5	0.3	0.65	0.38
zone1	MG	Gate	0.78	0.78	0.84	0.78	0.68	0.46
zone1	MG	T3	0.8	0.8	0.9	0.8	0.68	0.46
zone1	MG	T4	0.8	0.8	0.9	0.8	0.68	0.46
zone1	MG	T5	0.6	0.6	0.66	0.6	0.68	0.46
Gate			0.3	0.3	0.34	0.3	0.61	0.32
Zone2	Gate	T6	0.9	0.9	0.99	0.9	0.66	0.39

Table 7.6 Basic probabilities of successful intrusion for the activist

Typical	From	To	p_d	$\tilde{p}_a^{min}$	$\tilde{p}_a^{max}$	$\tilde{p}_a^{nominal}$	Coe_2	Coe_3
zone0		T1 or T2	0.95	0.9	0.97	0.95	0.68	0.45
MG			0.3	0.2	0.32	0.3	0.65	0.38
Dock			0.28	0.26	0.29	0.28	0.53	0.3
zone1	MG	Gate	0.78	0.7	0.8	0.78	0.68	0.46
zone1	MG	T3	0.8	0.72	0.8	0.8	0.68	0.46
zone1	MG	T4	0.8	0.72	0.8	0.8	0.68	0.46
zone1	Dock	Gate	0.78	0.7	0.78	0.78	0.68	0.46
zone1	Dock	T3	0.8	0.74	0.8	0.8	0.68	0.46
zone1	Dock	T4	0.8	0.78	0.8	0.8	0.68	0.46
Gate			0.2	0.15	0.21	0.2	0.61	0.32
Zone2	Gate	T6	0.9	0.8	0.9	0.9	0.66	0.39

the $\tilde{p}_a$ columns are the defender's estimation of the attacker's data. The $\tilde{p}_a^{min}$, $\tilde{p}_a^{max}$, and $\tilde{p}_a^{nominal}$ columns thus denote the defender minimal, maximal, and nominal estimation of the parameters. In a simultaneous game setting, the $\tilde{p}_a^{nominal}$ column is the attacker's own estimation and it is assumed to be known by the defender (cfr. the 'common knowledge' assumption). The $\tilde{p}_a^{min}$ and $\tilde{p}_a^{max}$ columns are not used in the simultaneous CPP game, since we only investigated sequential games with distribution-free uncertainties (Chap. 4). More explanation of the mutual knowledge of data is given in Sect. 2.1.5. The duration that the attacker stays in a zone affects his success probability. Generally speaking, in the same zone with the same security alert level (e.g., patrolling intensity), the longer the attacker stays in the zone, the more likely that he will be detected. Therefore, the "From" and "To" column are added to indicate different routes in each zone.

Probabilities in the "p_d" and "$\tilde{p}_a$" columns are estimated based on a lowest security alert level (e.g., a GREEN level or a level 1). Hereafter, a GREEN level is identical to level 1, a YELLOW level equals level 2, and a RED level is the same as level 3. If the "typical" has a higher SAL, such as a YELLOW (2) or a RED (3) level, extra tables should be provided by the security experts. In this study, for the sake of simplicity, we do not provide extra tables. Instead, the probabilities are assumed to

Fig. 7.3 The coefficients in Tables 7.5 and 7.6

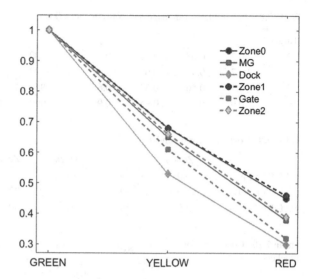

Table 7.7 Probabilities of damage and consequences (k€), for terrorist

Target	p_y	$\tilde{p}_y^{\,min}$	$\tilde{p}_y^{\,max}$	$\tilde{p}_y^{\,nominal}$	L	$\tilde{L}^{min}$	$\tilde{L}^{max}$	$\tilde{L}^{nominal}$
T1	0.1	0.1	0.12	0.1	1000	1100	1300	1200
T2	0.9	0.9	0.95	0.9	100	120	140	130
T3	0.7	0.7	0.8	0.7	300	240	260	250
T4	0.6	0.6	0.8	0.6	800	880	920	900
T5	0.9	0.9	0.99	0.9	2000	3000	3300	3000
T6	0.99	0.99	1	0.99	10,000	4900	5200	5000

decrease concavely [4]. The "Coe_2" and "Coe_3" columns show the decline coefficients, for a YELLOW level and for a RED level, respectively. Figure 7.3 demonstrates the concave property of these coefficients. Different lines denote different entrances or zones. Coefficients for the GREEN level are all set to be one, while coefficients for the YELLOW level and for the RED level are data from the "Coe_2" and "Coe_3" columns respectively. For example, in Table 7.5, $p_d(MG) = 0.3$ represents that, in a GREEN security alert level, the defender thinks that a terrorist with a VBIED scenario would have a probability of 0.3 to successfully pass the main entrance. While the $Coe_2(MG) = 0.65$ means that, if a YELLOW security alert level would be implemented at the main entrance, the defender thinks that a terrorist with a VBIED scenario would represent a probability of $0.3 \times 0.65 = 0.195$ to successfully pass the main entrance.

Tables 7.7 and 7.8 provide the estimations of conditional (in the condition that the attacker already arrived at the target) probabilities that an attack would be successfully executed and the accompanying estimated consequences/gains. The p_y column denotes the defender's estimation of the conditional probabilities that an attack would succeed under the condition that the attacker already reaches the target. The

Table 7.8 Probabilities of damage and consequences (k€), for activist

Target	p_y	$\tilde{p}_y^{\,min}$	$\tilde{p}_y^{\,max}$	$\tilde{p}_y^{\,nominal}$	L	$\tilde{L}^{\,min}$	$\tilde{L}^{\,max}$	$\tilde{L}^{\,nominal}$
T2	0.7	0.68	0.72	0.7	200	105	115	110
T3	0.7	0.4	0.6	0.5	300	280	310	300
T4	0.7	0.56	0.64	0.6	850	880	910	900
T6	0.9	0.85	0.95	0.9	1000	1800	2200	2000

Table 7.9 Costs (k€) for defender

	Zone0	MG	Dock	Zone1	Gate	Zone2
SAL:GREEN	40	20	20	20	20	20
SAL:YELLOW	60	30	25	30	25	30
SAL:RED	100	50	40	50	40	50

Table 7.10 Costs (k€) for attackers

Terrorist			Activist		
C_a^{min}	C_a^{max}	$C_a^{nominal}$	C_a^{min}	C_a^{max}	$C_a^{nominal}$
5	15	10	0.2	2	1

$\tilde{p}_y$ columns represent the attacker's estimation; the explanation of the three columns are similar to the columns in Tables 7.5 and 7.6. The L and $\tilde{L}$ columns denote the estimated losses and gains, for the defender and for the attacker respectively. Tables 7.9 and 7.10 further give the materialized defensive and attack costs respectively.

It is worth noting that the data provided in Tables 7.5, 7.6, 7.7, 7.8, 7.9, and 7.10 are all illustrative for this case study. If the CPP game would be used in industrial practice, the data should evidently be provided by security experts, for instance, by a company security team.

7.1.3   CPP Game Results

The game modelled for the case study is a 2-player game. The defender is determined, while the attacker can be either a terrorist or an activist. The prior probabilities of these two types of attackers are $p^t = (3/7, 4/7)$. The defender has $n = 3^{1+\sum_{r=1}^{2}(ent(r)+sub(r))} = 729$ pure strategies. The terrorist has $m^{terr} = 6$ pure strategies and the activist has $m^{acti} = 7$ pure strategies, as shown in Tables 7.3 and 7.4. If Formulas (3.7), (3.8), (4.2), (4.3), (4.20), (4.25), and (4.27) are filled in with data from tables shown in Sect. 7.1.2, the defender's payoff matrices U_d^{terr} and U_d^{acti}, the attacker's nominal payoff matrices U_a^{terr} and U_a^{acti}, the attacker's lower bound payoff matrices $\underline{U}_a^{terr}$ and $\underline{U}_a^{acti}$, the attacker's upper bound payoff matrices $\bar{U}_a^{terr}$ and $\bar{U}_a^{acti}$, and the coefficients matrices Ω^{terr} and Ω^{acti} can be obtained.

By employing the definitions and the algorithms proposed in this book, several solutions are investigated for the CPP game of the case study. For the sake of clarity, we first show the results of the CPP game of the case study by considering one type of attacker, in Sect. 7.1.3.1. Subsequently, we discuss the results of the game by considering multiple types of attackers (i.e., both the terrorist and the activist) in Sect. 7.1.3.2.

7.1.3.1 Single Attacker Type: The Activist

The Nash Equilibrium, the (Strong) Stackelberg Equilibrium, the Interval CPP Game solutions, the MoSICP solution, and the MiniMax solution are investigated for the case study by considering only the activist attacker. Results for the game only considering the terrorist attacker can be obtained analogously.

7.1.3.1.1 Nash Equilibrium

In the case study (actually, hereafter in the entire Sect. 7.1.3.1, the "case study" means the case study that only considers the activist and the defender), the defender and the activist are outguessing each other's pure strategy, and they have negatively correlated interests in the game. Therefore, there is no pure strategy Nash Equilibrium for the developed game. Table 7.11 illustrates the defender's best pure strategy responses (the second column) to each strategy of the activist (the first column), according to the defender's payoff matrix U_d^{acti}. The Activist's best response strategy to the defender's best response strategy is shown in the third column in the table. For instance, in the first row of Table 7.11, if the activist would play the strategy s_{e1}, then the defender's best response (a strategy that brings the player the highest payoff) is to play the strategy $2 \times 1 \times 1 \times 1 \times 1 \times 1$. Similarly, if the activist knows that the defender is going to play the strategy $2 \times 1 \times 1 \times 1 \times 1 \times 1$, then his best response is to play s_{e4}. Therefore, in Table 7.11, if in a row that the activist strategy (the first column) equals the activist's best response to the defender's best response (the third column), then it means that the activist's strategy and the defender's strategy in this

Table 7.11 Players' pure strategy best responses to their opponent's strategies

Activist strategy	Defender's best response	Activist's best response to the defender's best response
s_{e1}	$2 \times 1 \times 1 \times 1 \times 1 \times 1$	s_{e4}
s_{e2}	$1 \times 2 \times 1 \times 1 \times 1 \times 1$	s_{e5}
s_{e3}	$1 \times 1 \times 2 \times 1 \times 1 \times 1$	s_{e4}
s_{e4}	$1 \times 3 \times 1 \times 2 \times 1 \times 1$	s_{e5}
s_{e5}	$1 \times 1 \times 3 \times 2 \times 1 \times 1$	s_{e4}
s_{e6}	$1 \times 1 \times 1 \times 1 \times 2 \times 1$	s_{e4}
s_{e7}	$1 \times 1 \times 2 \times 1 \times 2 \times 1$	s_{e4}

Table 7.12 Mixed strategy NE

Pure strategy: Defender	Probability	Pure strategy: Activist	Probability
$2 \times 1 \times 1 \times 2 \times 1 \times 1$	0.6392	s_{e1}	0.3628
$2 \times 2 \times 1 \times 2 \times 1 \times 1$	0.2249	s_{e4}	0.4555
$2 \times 2 \times 2 \times 2 \times 1 \times 1$	0.1359	s_{e5}	0.1817

row are mutual best responses to each other, and furthermore, the row shows a pure strategy NE for the CPP game. There is no row in Table 7.11 that the activist strategy column equals the third column, which means that no pair of the players' strategies satisfy the mutual best responses condition, thus no pure Nash Equilibrium exists in the game.

By employing the Lemke-Howson algorithm [5] and taking U_a^{acti} and U_d^{acti} as inputs, we obtain one mixed strategy Nash Equilibrium for the game, as shown in Table 7.12.

The Nash Equilibrium indicates that the defender should always set the security alert levels (SAL) at Zone 0 and Zone 1_1 as level 2 ("YELLOW") while at the Gate and Zone 2_1 as level 1 ("GREEN"). The SAL at the Main Entrance (MG) should be set as "GREEN" by a probability of 63.92% and as "YELLOW" by a probability of $22.49\% + 13.59\% = 36.08\%$. The SAL at the Dock #1 should be set as "GREEN" by a probability of $63.92\% + 22.49\% = 86.41\%$ and as "YELLOW" by a probability of 13.59%. The activist would attack target T2 by a probability of 36.28% and would attack target T4 by a probability of 63.72% by passing perimeter 1 through the MG and through the Dock #1 by probabilities of 45.55% and of 18.17% respectively.

Table 7.13 further illustrates the probabilities that the activist would successfully reach the target, from the defender's and the attacker's point of view respectively. For instance, if the activist plays s_{e5}, then he has to pass Zone 0, Dock #1, and Zone 1_1, further if the defender plays a pure strategy $s_d = 2 \times 2 \times 2 \times 2 \times 1 \times 1$, then the SAL at Zone 0, the Dock #1, and Zone 1_1 are all "YELLOW". In this case, the probability that the activist will successfully reach target T4 can be calculated as $P = 0.95 * 0.68 * 0.28 * 0.53 * 0.8 * 0.68 = 0.0522$ (numbers are given in Table 7.6), resulting in the data shown in the bottom right cell in Table 7.13.

Table 7.14 further demonstrates defence costs and conditional expected losses and gains for each of the pure strategies that are played in the equilibrium. Note that the defence cost only depends on the defender's pure strategies while the conditional expected losses and gains only depend on the attacker's target. Hence, they can be given separately.

According to the results in Tables 7.12, 7.13, and 7.14, the defender obtains an equilibrium payoff of approximately[1] -242.0 k€ while the activist's equilibrium payoff equals 48.7 k€, as calculated by Formulas (7.1) and (7.2).

---

[1]In this Chapter, all the payoff values are rounded to the nearest tenth.

Table 7.13 Probability that the attacker can reach the target sucesfully

Def's Stg	Acti's Stg		
	s_{e1}	s_{e4}	s_{e5}
$2 \times 1 \times 1 \times 2 \times 1 \times 1$	$P : 0.646, \tilde{P} : 0.646$	$P : 0.1054, \tilde{P} : 0.1054$	$P : 0.0984, \tilde{P} : 0.0984$
$2 \times 2 \times 1 \times 2 \times 1 \times 1$	$P : 0.646, \tilde{P} : 0.646$	$P : 0.0685, \tilde{P} : 0.0685$	$P : 0.0984, \tilde{P} : 0.0984$
$2 \times 2 \times 2 \times 2 \times 1 \times 1$	$P : 0.646, \tilde{P} : 0.646$	$P : 0.0685, \tilde{P} : 0.0685$	$P : 0.0522, \tilde{P} : 0.0522$

Table 7.14 Defence cost and conditional expected losses and gains

Defender pure strategy	Defence cost (k€)	Activist pure strategy	Conditional expected loss and gain (k€)
$2 \times 1 \times 1 \times 2 \times 1 \times 1$	170	s_{e1}	$PL : 140, \tilde{PL} : 77$
$2 \times 2 \times 1 \times 2 \times 1 \times 1$	180	s_{e4}	$PL : 595, \tilde{PL} : 540$
$2 \times 2 \times 2 \times 2 \times 1 \times 1$	185	s_{e5}	$PL : 595, \tilde{PL} : 540$

$$u_d^* = [0.6392, 0.2249, 0.1359] \times u_{d\_temp} \times \begin{bmatrix} 0.3628 \\ 0.4555 \\ 0.1817 \end{bmatrix} \approx -242.0 \qquad (7.1)$$

$$u_a^* = [0.6392, 0.2249, 0.1359] \times u_{a\_temp} \times \begin{bmatrix} 0.3628 \\ 0.4555 \\ 0.1817 \end{bmatrix} \approx 48.7 \qquad (7.2)$$

$$u_{d\_temp} = - \begin{bmatrix} 0.646 & 0.1054 & 0.0984 \\ 0.646 & 0.0685 & 0.0984 \\ 0.646 & 0.0685 & 0.0522 \end{bmatrix} . \times \begin{bmatrix} 140 & 595 & 595 \\ 140 & 595 & 595 \\ 140 & 595 & 595 \end{bmatrix}$$
$$- \begin{bmatrix} 170 & 170 & 170 \\ 180 & 180 & 180 \\ 185 & 185 & 185 \end{bmatrix}$$

$$u_{a\_temp} = \begin{bmatrix} 0.646 & 0.1054 & 0.0984 \\ 0.646 & 0.0685 & 0.0984 \\ 0.646 & 0.0685 & 0.0522 \end{bmatrix} . \times \begin{bmatrix} 77 & 540 & 540 \\ 77 & 540 & 540 \\ 77 & 540 & 540 \end{bmatrix} - \begin{bmatrix} 1 & 1 & 1 \\ 1 & 1 & 1 \\ 1 & 1 & 1 \end{bmatrix}$$

7.1.3.1.2 Stackelberg Equilibrium

The pure strategy Stackelberg Equilibrium of the game is that: $\bar{s}_d = 2 \times 1 \times 1 \times 2 \times 1 \times 1$ and $\bar{s}_a = s_{e4}$. Figure 7.4 shows the defender's payoffs and the activist's best responses to each of the defender's pure strategies. The x axis denotes the defender's pure strategies, and concerns in total a number of $n = 729$ points. The right-hand side y axis together with the 'o' (and brown) points represent the activist's best responses to the defender's pure strategies. The left-hand side y axis together with the '+' (and blue) points denote the defender's payoffs in case that the defender plays a pure strategy and the activist plays a best response. The square (and red) point on top of the figure denotes the highest payoff that the defender can obtain by playing pure strategies in a Stackelberg game. The square (and red) point's corresponding defender pure strategy (on the x axis) and activist's best response (on the right-hand side y axis) are the defender and the activist's strategies of a pure strategy Stackelberg Equilibrium. The corresponding strategies are $\bar{s}_d = 2 \times 1 \times 1 \times 2 \times 1 \times 1$ and $\bar{s}_a = s_{e4}$, and the corresponding defender's Stackelberg Equilibrium payoff is -232.7 k€.

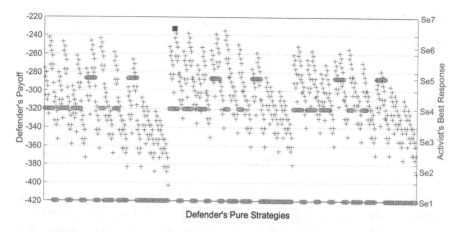

Fig. 7.4 Defender's payoff by responding with different strategies

Table 7.15 SSE strategies of the game

Defender pure strategy	Probability
$2 \times 1 \times 1 \times 2 \times 1 \times 1$	0.6392
$2 \times 2 \times 1 \times 2 \times 1 \times 1$	0.2249
$2 \times 2 \times 2 \times 2 \times 1 \times 1$	0.1359
Activist's pure strategy	Probability
s_{e4}	1

Table 7.16 Players' payoffs when the activist play different strategies and the defender plays her SE strategy

Activist's strategy	Activist's payoff	Defender's payoff
s_{e1}	**48.7420**	**−264.7271**
s_{e2}	12.8172	−193.6312
s_{e3}	12.8172	−193.6312
s_{e4}	**48.7420**	**−229.0954**
s_{e5}	**48.7420**	**−229.0954**
s_{e6}	28.0991	−188.8367
s_{e7}	28.0991	−188.8367

Table 7.15 shows the mixed strategy Stackelberg Equilibrium for the game, obtained by employing the MultiLP algorithm [6] and taking U_a^{acti} and U_d^{acti} as input information. It is interesting to notice that the defender's Stackelberg Equilibrium strategy is identical to her Nash Equilibrium strategy (shown in Table 7.12). The fact that these equilibria are the same stems from the observation that when the defender plays her Nash (Stackelberg) Equilibrium strategy, the activist has the same payoffs by responding with strategies s_{e1}, s_{e4}, and s_{e5}, as shown in the second column of Table 7.16. Moreover, this payoff is higher than the payoffs obtained if responding with other strategies. Furthermore, the third column of Table 7.16 shows the defender's payoff when the activist responds different strategies. It reveals that though being indifferent among strategies s_{e1}, s_{e4}, and s_{e5}, the activist plays a

Table 7.17 Defender's
optimal strategy from the
IBGS algorithm

Defender pure strategy	Probability
$2 \times 1 \times 1 \times 1 \times 1 \times 1$	0.5817
$2 \times 1 \times 1 \times 2 \times 1 \times 1$	0.2426
$2 \times 2 \times 1 \times 1 \times 1 \times 1$	0.1757

Table 7.18 Defender's
optimal strategy from the
ICGS algorithm

Index	Defender's pure strategy	Probability
$s_{d-ICGS-1}$	$2 \times 1 \times 1 \times 1 \times 1 \times 1$	0.3238
$s_{d-ICGS-2}$	$2 \times 1 \times 1 \times 2 \times 1 \times 1$	0.5173
$s_{d-ICGS-3}$	$2 \times 2 \times 1 \times 1 \times 1 \times 1$	0.1589

strategy that benefits the defender, and otherwise if the activist plays s_{e1}, the
defender's payoff would significantly decrease to -264.7 k€. The activist (the
game follower)'s behaviour of choosing a favour strategy for the defender (the
game leader) is called the "breaking-tie" assumption, as we discussed in Sect.
2.2.3. Therefore, the defender's expected payoff resulting from the mixed strategy
Stackelberg Equilibrium is -229.1 k€ and the activist's payoff equals 48.7 k€.

7.1.3.1.3 Interval CPP Game Solution

Tables 7.17 and 7.18 give the defender's optimal solution of the game from the
Interval Bi-Matrix Game Solver (IBGS algorithm) and from the Interval CPP Game
Solver (ICGS algorithm) respectively, in case that she has distribution-free uncer-
tainties on the activist's parameters. U_d^{acti}, U_a^{acti}, $\underline{U}_a^{acti}$, $\bar{U}_a^{acti}$, and Ω^{acti} are used as
input information for these two algorithms.

The activist's payoffs by responding with different strategies to the defender's
IBGS optimal strategy and to the defender's ICGS optimal strategy are shown on the
left-hand side panel and on the right-hand panel of Fig. 7.5, respectively. The x axis
of Fig. 7.5 denotes the activist's pure strategies, the y axis denotes the activist's
payoff. The bold red dots denote the activist's payoff by responding with different
pure strategies, without considering distribution-free uncertainties. Conversely, if
the distribution-free uncertainties are considered, then the activist's payoffs by
responding with different strategies to the defender's optimal strategy cannot be
obtained directly and only the ranges can be calculated, as shown in the figure by
vertical lines. The horizontal lines represents the activist's maximal lower bound
payoffs.

The left-hand side panel of Fig. 7.5 shows that if the defender plays her IBGS
optimal strategy and if there would be no uncertainties, then the activist's best
response would be either s_{e4} or s_{e5}. When uncertainties are considered, then the
defender knows that strategy s_{e5} can deliver the activist the highest lower bound
payoff, being 54.4 k€. Furthermore, the upper bound payoff of strategies s_{e4}, s_{e5}, and
s_{e6} are higher than 54.4 k€, thus these three strategies all have the possibilities of
being the attacker's best response. The upper bound payoff of strategy s_{e1} equals

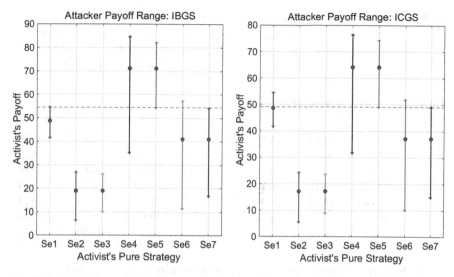

Fig. 7.5 Attacker's payoff range

Table 7.19 Activist's payoff differences

Payoff differences	Result involving the shared parameters	Comments
$\Delta_{se5\,-\,se1}$	$0.5971 \cdot \tilde{P}_0^{\,z}$	>0
$\Delta_{se5\,-\,se4}$	$-0.0302 \cdot \tilde{P}_0^{\,z} \cdot \tilde{p}_y \cdot \tilde{L}$	<0, the $\tilde{p}_y \cdot \tilde{L}$ denotes the conditional expected loss of target T4
$\Delta_{se5\,-\,se6}$	$4.6370 \cdot \tilde{P}_0^{\,z}$	>0

exactly 54.4 k€, and it is excluded from the activist's possible best response set, as we explained in Sect. 4.3. The defender's payoffs would be -243.6 k€, -243.6 k€, and -185.3 k€ respectively, if the activist responds with s_{e4}, s_{e5}, or s_{e6}. The defender conservatively treats the worst case as the real case, thus the defender has an expected payoff from the IBGS solution of -243.6 k€.

The right-hand side panel of Fig. 7.5 reveals that if the defender plays her ICGS optimal solution, then s_{e5} would deliver the activist the highest lower bound payoff and strategies s_{e1}, s_{e4}, s_{e5}, and s_{e6} are all possible best responses (according to the criteria used in the IBGS algorithm). However, strategies s_{e1}, s_{e4}, and s_{e6} share some parameters (e.g., $\tilde{P}_0^{\,z}$) with strategy s_{e5}. The activist's payoff differences $\Delta_{se5\,-\,se1}$, $\Delta_{se5\,-\,se4}$, and $\Delta_{se5\,-\,se6}$ indicate that strategies s_{e1} and s_{e6} are always worse than strategy s_{e5}, as shown in Table 7.19. Therefore, the ICGS algorithm leads to s_{e4} and s_{e5} to be the activist's possible best response strategies. The defender's expected payoff from the ICGS solution equals -238.6 k€, being higher than the defender's expected payoff from the IBGS solution.

Table 7.20 Defender's MoSICP strategy

Index	Defender's pure strategy	Probability
$s_{d-MoSICP-1}$	$2 \times 1 \times 1 \times 1 \times 1 \times 1$	0.5292
$s_{d-MoSICP-2}$	$2 \times 1 \times 1 \times 2 \times 1 \times 1$	0.3754
$s_{d-MoSICP-3}$	$2 \times 1 \times 2 \times 1 \times 1 \times 1$	0.0954

Table 7.21 Δ matrix of the MoSICP

(i,j)	1	2	3	4	5	6	7
1	0	**18.06**	**21.64**	−39.43	−27.33	−11.64	−2.42
2	−45.25	0	−8.33	−71.55	−65.06	−43.71	−39.85
3	−42.75	−7.92	0	−74.66	−61.18	−46.82	−35.98
4	−13.43	**22.19**	**16.16**	0	−25.91	**4.18**	−8.14
5	**0.00**	25.43	**31.30**	−21.48	0	−3.64	**10.14**
6	−39.70	−3.77	−8.86	−62.66	−59.50	0	−17.40
7	−36.21	−9.23	−3.24	−68.13	−53.89	−19.09	0

The following explains how $\Delta_{se5-se4}$ can be obtained. $\Delta_{se5-se1}$ and $\Delta_{se5-se6}$ can be calculated analogously, as theoretically explained in Sect. 4.4. Numbers used in the following formulas are derived from Table 7.6.

$$u_a^{min}(s_{e5}, s_{d-ICGS-1}) = \tilde{P}_0^z \cdot 0.26 \cdot 0.78 \cdot \tilde{p}_y \cdot \tilde{L} - C_a$$
$$u_a^{min}(s_{e5}, s_{d-ICGS-2}) = \tilde{P}_0^z \cdot 0.26 \cdot 0.78 \cdot 0.68 \cdot \tilde{p}_y \cdot \tilde{L} - C_a$$
$$u_a^{min}(s_{e5}, s_{d-ICGS-3}) = \tilde{P}_0^z \cdot 0.26 \cdot 0.78 \cdot \tilde{p}_y \cdot \tilde{L} - C_a$$
$$u_a^{max}(s_{e4}, s_{d-ICGS-1}) = \tilde{P}_0^z \cdot 0.32 \cdot 0.8 \cdot \tilde{p}_y \cdot \tilde{L} - C_a$$
$$u_a^{max}(s_{e4}, s_{d-ICGS-2}) = \tilde{P}_0^z \cdot 0.32 \cdot 0.8 \cdot 0.68 \cdot \tilde{p}_y \cdot \tilde{L} - C_a$$
$$u_a^{max}(s_{e4}, s_{d-ICGS-3}) = \tilde{P}_0^z \cdot 0.32 \cdot 0.65 \cdot 0.8 \cdot \tilde{p}_y \cdot \tilde{L} - C_a$$

$$\begin{aligned}
\Delta_{se5-se4} &= 0.3238 \cdot \left(u_a^{min}(s_{e5}, s_{d-ICGS-1}) - u_a^{max}(s_{e4}, s_{d-ICGS-1})\right) \\
&+ 0.5173 \cdot \left(u_a^{min}(s_{e5}, s_{d-ICGS-2}) - u_a^{max}(s_{e4}, s_{d-ICGS-2})\right) \\
&+ 0.1589 \cdot \left(u_a^{min}(s_{e5}, s_{d-ICGS-3}) - u_a^{max}(s_{e4}, s_{d-ICGS-3})\right) \\
&= -0.0302 \cdot \tilde{P}_0^z \cdot \tilde{p}_y \cdot \tilde{L}
\end{aligned}$$

7.1.3.1.4 MoSICP Solution

The defender's optimal strategy from the MoSICP solution is given in Table 7.20. U_d^{acti} and Ω^{acti} are the inputs for the algorithm proposed in Sect. 5.3.

Table 7.21 shows the Δ matrix. According to the definition of MoSICP in Sect. 5.3, we know that:

$$E(y) = \{(1,2), (1,3), (4,2), (4,3), (4,6), (5,2), (5,3), (5,7)\} \quad (7.3)$$
$$Q(y) = \{x \in X | x_1 \geq x_2, x_3; x_4 \geq x_2, x_3, x_6; x_5 \geq x_1, x_2, x_3, x_7\} \quad (7.4)$$

Table 7.22 Attacker's strategy in the MoSICP

Activist's strategy	Activist's payoff
s_{e1}	**0.1892**
s_{e2}	0
s_{e3}	0
s_{e4}	**0.6216**
s_{e5}	**0.1892**
s_{e6}	0
s_{e7}	0

Table 7.23 Defender's MiniMax solution strategy

Defender's pure strategy	Probability
$2 \times 1 \times 1 \times 1 \times 1 \times 1$	0.9440
$2 \times 2 \times 1 \times 1 \times 1 \times 1$	0.0560

Table 7.24 Player's payoff when the activist plays different strategies

Activist's response	Defender's payoff	Activist's payoff
s_{e1}	**−251.0002**	48.7420
s_{e2}	−192.4802	21.8000
s_{e3}	−190.9481	20.7056
s_{e4}	**−251.0002**	**81.0800**
s_{e5}	−246.6591	77.1402
s_{e6}	−184.5686	47.0168
s_{e7}	−183.4162	44.7120

Table 7.22 further shows the activist's strategy in the MoSICP solution. In the work of Jiang et al. [7] (Lemma 1) and Nguyen et al. [8], it is proven that "each action that is played with positive probability in the attacker's mixed strategy is played with equal probability". The above statement means that, any actions that have the probability of being played (i.e., strategies s_{e1}, s_{e4}, and s_{e5} in our case) should have the same probability of being played (i.e., the probabilities shown in the right-hand column should be 0.33, 0.33 and 0.33 for strategies s_{e1}, s_{e4}, and s_{e5} respectively). The mixed strategy shown in Table 7.22 does not fit this lemma. In their work, attacker's parameter uncertainties are not modelled, thus a full order theory is employed. The full order theory results in the equality of probabilities of different played actions. However, in the MoSICP, due to the existence of the distribution-free uncertainties, it is not necessary that all the played actions will be played by the same probabilities. To be more specific, we will notice that set Q (shown in Formula (7.4)) does not show any relationships between x_4 and x_1, or x_4 and x_5.

The defender's expected payoff from the MoSICP solution is −245.4 k€.

7.1.3.1.5 MiniMax Solution

Table 7.23 shows the defender's optimal strategy from the MiniMax solution. Table 7.24 further illustrates the defender's and the activist's payoff with regard to

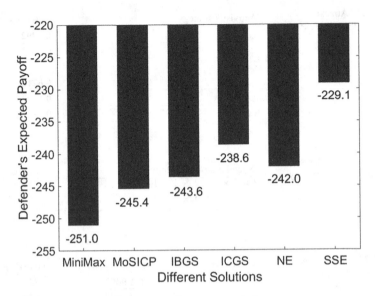

Fig. 7.6 Defender's expected payoff from different game solutions

different activist's responses to the defender's optimal strategy. The third column shows that the activist's best response should be s_{e4}, bringing the activist a payoff of 81.1 k€. However the second column shows that a MiniMax activist would also play strategy s_{e1}, since s_{e1} minimizes the defender's payoff as well. The second column of Table 7.24 indicates that the defender's optimal payoff from the MiniMax solution would be -251.0 k€.

7.1.3.1.6 (Inter-) Comparison of Different Solutions

Figure 7.6 shows the defender's expected payoffs (y axis) from different solutions (x axis). The MiniMax represents the result from an activist who aims at minimizing the defender's payoff; the MoSICP denotes the result from an activist who would play pure strategies that have higher payoffs with higher probabilities and the defender has distribution-free uncertainties on the activist's parameters; the IBGS and the ICGS denote the results from a rational activist but where the defender has distribution-free uncertainties on the activist's parameters; the NE and SSE represent the results for a game played with a rational activist and complete information, for a simultaneous moving game and for a sequential moving game, respectively.

Figure 7.6 reveals that if the defender has complete information of the game while the activist has perfect information of the game (i.e., an SSE solution), the defender will have the maximal payoff, being -229.1 k€. The defender's uncertainties on the activist would reduce her expected payoff (i) to -238.6 k€ in case of only having distribution-free uncertainties on parameters (i.e., an ICGS solution), (ii) to -245.4 k € in case of having uncertainties on both the activist's parameters and rationality

Fig. 7.7 Robustness of different solutions

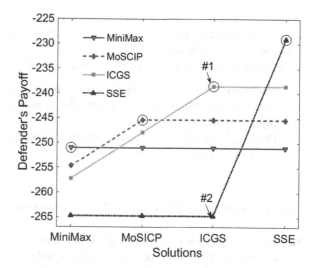

(i.e., an MoSICP solution), and (iii) to -251.0 k€ in case of having no information about the attacker at all (i.e., an MiniMax solution). Comparison of the results of the NE solution and the SSE solution reveals the "first-mover advantage" for the defender. The NE brings the defender a payoff of -242.0 k€, being less than the payoff that the defender obtains from the SSE. The payoff differences between the IBGS and the ICGS proves the effectiveness of the algorithm ICGS by taking into account the parameter coupling problem of the CPP game.

Figure 7.7 demonstrates the robustness of different solutions to the case study game. Different lines in the figure denote the robustness of different solutions, i.e., as shown in the legend, the MiniMax solution, the MoSICP solution, the ICGS solution, and the Strong Stackelberg Equilibrium. The y axis denotes the defender's payoff. The x axis represents the real situation of the activist's rationality or the defender's information. For instance, the SSE line (in purple colour) denotes that the defender believes that she has complete information of the game and the activist is a rational player. Therefore, the defender plays her SSE strategy (as shown in Table 7.15). However, the real situation may not be the same one as that which the defender believes. The four points in the SSE line represent the defender's real payoff in case of a corresponding real situation. The point "#2", for example, means that the defender's information of the attacker is incorrect and distribution-free uncertainties actually exist (as assumed in the interval CPP game), and the defender thus has a lower payoff, being -264.7 k€. Points with a circle (e.g., the point "#1") denote the situation that the assumptions of the line are satisfied.

Figure 7.7 reveals that solutions with less strict assumptions are more robust than others. The MiniMax solution, which is the most conservative solution for the defender and does not require any assumptions on the activist's behaviour and on the defender's knowledge of the activist's parameters, ensures a payoff of -251.0 k€ to the defender. Other solutions, though promising higher payoffs to the defender if the required assumptions are satisfied, may pull the defender to a quite worse situation if her assumptions would not hold and would not be true. Furthermore, the more strict assumptions the solution need, the worse the result will be if the assumptions would be false.

Two conclusions can be drawn from Figs. 7.6 and 7.7:

(i) It is important for the defender to collect useful security data. Sufficient intelligence (data) would support the defender to play a better strategy to obtain a higher payoff.
(ii) The reliability of the collected intelligence (data) is important. Fake/false information is worse than no information.

7.1.3.2 Multiple Attacker Types: The Terrorist and the Activist

Results of the game involving multiple types of attackers, namely, the terrorist and the activist, are given. The Bayesian Nash Equilibrium, the Bayesian Stackelberg Equilibrium, the Interval CPP game solutions, the robust solution with epsilon-optimal attacker, and the MiniMax solution are all calculated. A thorough comparison of these different solutions is also given.

7.1.3.2.1 Bayesian Nash Equilibrium

Table 7.25 shows the defender's and the attackers' Bayesian Nash Equilibrium (BNE) strategy. Tables 7.25 and 7.26 demonstrate that the attackers are playing thier best response strategies.

Figure 7.8 illustrates the defender's payoff by responding with different pure strategies to the attackers' BNE strategy. The x axis denotes different defender pure strategies, while the blue triangles represent the defender's payoff (y axis) when she plays the corresponding pure strategy (x axis). Therefore, there are in total 729 triangles in the figure. The two triangles surrounded by red circles denote the best response strategies, and the corresponding defender pure strategies are $2 \times 2 \times 1 \times 2 \times 1 \times 1$ and $2 \times 2 \times 2 \times 2 \times 1 \times 1$ respectively, while the corresponding defender's payoff is -265.5 k€. Recalling the defender's BNE

Table 7.25 BNE strategies

Index	Defender pure strategy	Probability
$s_{d-BNE-1}$	$2 \times 2 \times 1 \times 2 \times 1 \times 1$	0.8641
$s_{d-BNE-2}$	$2 \times 2 \times 2 \times 2 \times 1 \times 1$	0.1359
Terrorist pure strategy	Probability	
s_{v5}	1	
Activist pure strategy	Probability	
s_{e1}	0.6820	
s_{e5}	0.3180	

Table 7.26 Attackers' payoff by responding with different strategies to the defender's BNE strategy

Terrorist pure strategy	Payoff	Activist pure strategy	Payoff
s_{v1}	67.5200	s_{e1}	**48.7420**
s_{v2}	65.5820	s_{e2}	9.2792
s_{v3}	1.9923	s_{e3}	12.8172
s_{v4}	27.0049	s_{e4}	36.0049
s_{v5}	**128.7686**	s_{e5}	**48.7420**
s_{v6}	79.2976	s_{e6}	20.6479
		s_{e7}	28.0991

Fig. 7.8 Defender's payoffs by responding with pure strategies to the attackers' BNE strategies

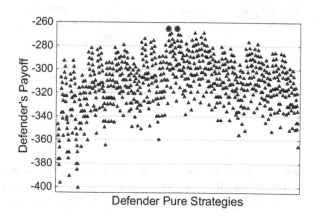

strategy given in Table 7.25, it is shown that the defender is also playing her best response strategies.

In conclusion, in the BNE, the defender and the attackers are all playing their best response strategies to their opponents' strategies.

7.1.3.2.2 Bayesian Stackelberg Equilibrium

The defender's Bayesian Stackelberg Equilibrium strategy we obtained for the game is the same as her Bayesian Nash Equilibrium strategy as shown in Table 7.25. The defender's expected payoff from the BSE is -251.6 k€, being higher than her expected payoff from the BNE, which is -265.5 k€.

By playing the same strategy, the defender receives different expected payoffs from the BNE and the BSE. This is the result of the "breaking-tie" assumption (see Sect. 2.2.3 for more information). As shown in Table 7.25, the activist has the same corresponding payoff (that is, 48.7 k€) by both responding strategies s_{e1} or s_{e5}. On the contrary, the defender's payoff would be -271.1 k€ and -235.5 k€ respectively, if the activist plays s_{e1} or s_{e5}. In the BNE solution, the activist is assumed not to be able to know the defender's strategy when he moves. Therefore, the activist randomly (but strategically) plays both strategy s_{e1} and s_{e5}. Conversely, in the BSE

Table 7.27 Slightly modified BSE strategies

Defender's BSE strategy

Index	Defender pure strategy	Probability
$s_{d-BSE-1}$	$2 \times 2 \times 1 \times 2 \times 1 \times 1$	0.8681
$s_{d-BSE-2}$	$2 \times 2 \times 2 \times 2 \times 1 \times 1$	0.1319

Terrorist's BSE strategy		Activist's BSE strategy	
Index	Probability	Index	Probability
s_{v5}	1	s_{e5}	1

Table 7.28 Attackers' payoff by responding with different strategies to the defender's modified BSE strategy

Terrorist pure strategy	Terrorist payoff	Defender payoff	Activist pure strategy	Activist payoff	Defender payoff
s_{v1}	67.5200	−245.2595	s_{e1}	48.7420	−271.0995
s_{v2}	65.5820	−238.7995	s_{e2}	9.2792	−195.0503
s_{v3}	1.9923	−195.0503	s_{e3}	12.8450	−200.0422
s_{v4}	27.0049	−213.5528	s_{e4}	36.0049	−221.4335
s_{v5}	**128.7686**	−273.1719	s_{e5}	**48.8420**	−235.5772
s_{v6}	79.2976	−359.2546	s_{e6}	20.6479	−191.4834
			s_{e7}	28.1576	−195.2381

solution, the activist knows the defender's strategy when he moves. Therefore, the activist knows that both strategies s_{e1} and s_{e5} will bring himself the best payoff and the "breaking-tie" assumption requires the activist to choose s_{e1} or s_{e5} preferable for the defender, resulting in strategy s_{e5} being chosen as the activist's best response.

The MovLib algorithm is applied on the BSE to relax the "breaking-tie" assumption, by slightly modifying the players' strategies, as discussed in Sect. 3.3.4. The ε is set as 0.1 for both the terrorist and the activist. The modified BSE is shown in Table 7.27. Table 7.28 further shows the attackers' payoffs by responding with different pure strategies to the defender's modified BSE strategy. It is shown that s_{v5} and s_{e5} are the unique best response to the terrorist and to the activist respectively. The defender's expected payoff from the modified BSE is only 0.0429 k€ less than the BSE payoff, being −251.7 k€,[2] while the "breaking-tie" assumption is no longer required.

7.1.3.2.3 Interval CPP Game Solution

Table 7.29 shows the defender's solutions of the Interval CPP game. Figure 7.9 illustrates the attackers' payoffs by responding with different pure strategies. The x and y axis, the vertical lines, the horizontal lines, and the red circles are the same as those we defined in Fig. 7.5 in Sect. 7.1.3.1.3. The two sub-figures on the top

---

[2]It is actually from −251.6466 k€ to −251.6895 k€.

Table 7.29 Defender's optimal strategy for the Interval Game

IBGS algorithm	
Defender pure strategy	Probability
$2 \times 2 \times 1 \times 1 \times 3 \times 1$	0.0542
$2 \times 2 \times 1 \times 2 \times 2 \times 1$	0.2426
$2 \times 3 \times 1 \times 1 \times 2 \times 1$	0.7032
ICGS algorithm	
Defender pure strategy	Probability
$2 \times 2 \times 1 \times 2 \times 1 \times 1$	0.5173
$2 \times 3 \times 1 \times 1 \times 1 \times 1$	0.4827

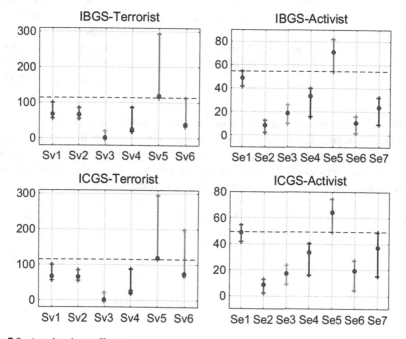

Fig. 7.9 Attackers' payoff range

(bottom) are the terrorist and the activist's payoffs to the defender's optimal solution from the IBGS (ICGS) algorithm.

Figure 7.9 shows that if the defender plays her IBGS strategy, strategies s_{v5} and s_{e5} have the highest lower bound payoffs for the terrorist and for the activist respectively. Moreover, the upper bound values of all other strategies are lower than the lower bound payoffs of these two strategies. Therefore, the terrorist and the activist's possible best responses are s_{v5} and s_{e5} respectively. If the defender plays her ICGS strategy, strategies s_{v5} and s_{e5} also have the highest lower bound payoffs for the terrorist and for the activist respectively. Furthermore, the terrorist's strategy s_{v6} and the activist's strategy s_{e1} have higher upper bound payoffs than the lower bound payoffs of strategies s_{v5} and s_{e5} respectively. Consequently, according to the

Table 7.30 Defender's
optimal strategy to the
epsilon-optimal attackers

Defender pure strategy	Probability
$2 \times 2 \times 1 \times 2 \times 1 \times 1$	0.9042
$2 \times 2 \times 2 \times 2 \times 1 \times 1$	0.0958

Table 7.31 Attackers' payoff by responding with different strategies to the defender's strategy from the epsilon-optimal solution

Terrorist pure strategy	Terrorist payoff	Defender payoff	Activist pure strategy	Activist payoff	Defender's payoff
s_{v1}	67.5200	−245.0792	s_{e1}	48.7420	−270.9192
s_{v2}	65.5820	−238.6192	s_{e2}	9.2792	−194.8700
s_{v3}	1.9923	−194.8700	s_{e3}	13.0950	−200.2122
s_{v4}	27.0049	−213.3725	s_{e4}	36.0049	−221.2531
s_{v5}	**128.7686**	−272.9915	s_{e5}	**49.7420**	−236.3893
s_{v6}	79.2976	−359.0743	s_{e6}	20.6479	−191.3031
			s_{e7}	28.6841	−195.3212

IBGS algorithm, both the terrorist and the activist would have two possible best response strategies. Conversely, the ICGS algorithm indicates that both attackers have only one possible best response. This is the result of the parameter coupling problem, and Formulas (7.5) and (7.6) explain why s_{v6} and s_{e1} should be excluded from the terrorist's and the activist's possible best response set.

$$\Delta^{terr}_{sv5-sv6} = 52.67 \cdot \tilde{P}^z_0(SAL:2) \cdot \tilde{P}_{MG}(SAL:2)\$\$ \qquad\qquad +72.28$$
$$\cdot \tilde{P}^z_0(SAL:2) \cdot \tilde{P}_{MG}(SAL:3)$$
$$> 0 \qquad\qquad\qquad\qquad\qquad\qquad\qquad\qquad\qquad (7.5)$$

$$\Delta^{acti}_{se5-se1} = 0.60 \cdot \tilde{P}^z_0(SAL:2) > 0 \qquad\qquad\qquad (7.6)$$

7.1.3.2.4   Robust Solution with Epsilon-Optimal Attacker

Table 7.30 shows the defender's optimal strategy from the robust solution with epsilon-optimal attackers. The epsilon is set as $\varepsilon = 1$ for both attackers. Table 7.31 illustrates the attackers' payoffs by responding with different strategies to the defender's strategy shown in Table 7.30. The terrorist will have the highest payoff by playing strategy s_{v5}, being 128.8 k€, while the activist's highest payoff strategy is s_{e5}. The attackers would deviate from their best response strategies, and the range of the deviation is set as $\varepsilon = 1$, thus any strategy that has a payoff 1.0 k€ less than the attackers' payoff from their best response strategies can be their possible best responses. Table 7.31 shows that the terrorist's possible response would only be s_{v5} and the activist's possible response would only be s_{e5}. Note that the activist's

Table 7.32 Defender's optimal strategy from the MiniMax solution

Defender pure strategy	Probability
$2 \times 2 \times 1 \times 2 \times 2 \times 1$	0.5564
$2 \times 2 \times 1 \times 2 \times 2 \times 2$	0.4436

Table 7.33 Players' payoffs when the attackers respond different strategies to the defender's optimal strategy

Terrorist pure strategy	Terrorist payoff	Defender payoff	Activist pure strategy	Activist payoff	Defender's payoff
s_{v1}	67.5200	−254.0358	s_{e1}	48.7420	**−279.8758**
s_{v2}	65.5820	−247.5758	s_{e2}	9.2792	−203.8267
s_{v3}	1.9923	−203.8267	s_{e3}	13.7598	−210.0996
s_{v4}	27.0049	−222.3291	s_{e4}	36.0049	−230.2098
s_{v5}	128.7686	**−281.9482**	s_{e5}	52.1353	−247.9831
s_{v6}	36.2562	−281.9482	s_{e6}	10.2136	−195.0427
			s_{e7}	15.1016	−197.4867

payoff by playing s_{e1} is exactly 1.0 k€ less than his payoff by playing s_{e5}, thus s_{e1} is excluded from the activist's possible best responses.

The defender's expected payoff by playing her robust solution as shown in Table 7.30 is therefore −252.1 k€, as calculated by Formula (7.7). Instead, if the defender does not take the attacker's bounded rationality into consideration and plays her Bayesian Stackelberg Equilibrium strategy, as shown in Table 7.27, then the activist may response s_{e1} as well as s_{e5} (though the terrorist still only responds with s_{v5}), resulting in the defender obtaining a quite low payoff, being −272.0 k€. The calculation result of this worst payoff is obtained by using Formula (7.8), and the numbers are taken from Table 7.28.

$$-272.9915 \times \frac{3}{7} - 236.3893 \times \frac{4}{7} = -252.0760. \tag{7.7}$$

$$-273.1719 \times \frac{3}{7} - 271.0995 \times \frac{4}{7} = -271.9877. \tag{7.8}$$

7.1.3.2.5 MiniMax Solution

Table 7.32 shows the defender's MiniMax strategy and Table 7.33 illustrates the players' payoffs when the attackers respond different pure strategies to the defender's MiniMax strategy.

The defender's worst expected payoff from the MiniMax solution is therefore −280.8 k€, as calculated by Formula (7.9).

$$-281.9482 \times \frac{3}{7} - 279.8758 \times \frac{4}{7} = -280.7640. \tag{7.9}$$

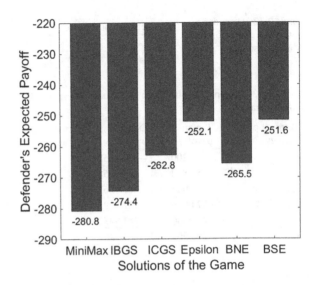

Fig. 7.10 Defender's expected payoffs from different solutions, considering multiple types of attackers

7.1.3.2.6   (Inter-) Comparison of Different Solutions

Figure 7.10 shows the defender's expected payoffs from different solutions of the CPP game, considering two types of attackers. The result is similar to the result we show in Sect. 7.1.3.1.6 where only one type of attacker is considered.

The first conclusion is that, the increase of the defender's uncertainties on the attackers would reduce the defender's expected payoff. The defender has an expected payoff of −251.6 k€ from the BSE solution, in which the defender is assumed to have complete information of the attackers and the attackers are rational players, while the defender's expected payoff is as low as −280.8 k€ from the MiniMax solution, in which the defender does not need any information of the attacker.

The second conclusion is that, in an idealistic situation, which means that both the defender and the attacker have complete information of the game and are rational players, a sequential moving game is better than a simultaneous moving game for the defender. The defender has a payoff of −265.5 k€ from the Bayesian Nash Equilibrium, being 13.9 k€ lower than her payoff from the Bayesian Stackelberg Equilibrium.

The third conclusion is that, in a multiple types of attackers situation and the defender has interval uncertainties on the attackers' parameters, the ICGS algorithm is more efficient for the defender than the IBGS algorithm. As shown in Fig. 7.10, in the same setting, the ICGS brings the defender a payoff of −262.8 k€ while the IBGS brings the defender a payoff of only −274.4 k€.

Figure 7.11 further demonstrates our first conclusion by showing the results of the sensitivity analysis. The y axis of the figure denotes the defender's expected optimal payoff. The two horizontal lines represent the defender's payoffs from the Bayesian Stackelberg Equlibrium (the line on the top) and from the MiniMax solution.

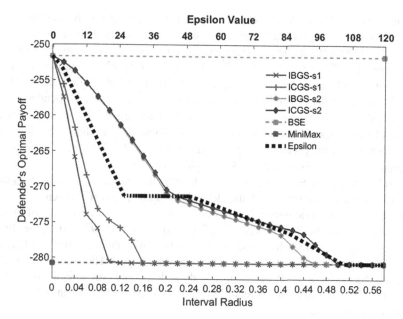

Fig. 7.11 Sensitivity analysis (of the epsilon value in the robust solution and of the interval radius in the interval game solution)

The black bold dot line together with the x axis on top show the result of the sensitivity analysis of the robust solution with epsilon-optimal attackers. The x axis denotes the value of epsilon, which can be interpreted as a measurement of the attacker's rationality: the higher the epsilon is, the less rational the attacker is. The black bold dot line reveals that the defender's expected payoff is declining when the attacker is becoming more and more irrational. At the most left-hand side, we have that epsilon equals zero, and the defender's expected payoff is as high as her BSE payoff. At the right-hand side, when epsilon equals 106 (see the reality meaning of this number in Sect. 5.2.1), the defender's expected payoff becomes as low as her MiniMax payoff. Therefore, if the defender has the confidence that the attackers' decision tolerance is less than 106, then it is useful and necessary to estimate the attacker's real tolerance value (i.e., epsilon). Otherwise, if the defender estimates that the attackers' decision tolerance is higher than 106, then it is not necessary anymore to know the value of the attacker's real tolerance, instead, the defender can play her MiniMax strategy directly.

Furthermore, we study how the interval uncertainties on the attackers would affect the defender's optimal payoffs, and results are shown by the four curves (together with the x axis on the bottom) with legend as "IBGS-s1", "ICGS-s1", "IBGS-s2", and "ICGS-s2". Two experiments are defined, namely, s1, in which the defender has interval uncertainties on all the attackers' parameters, and s2, in which the defender only has interval uncertainties on the attackers' monetary parameters. In both experiments, the defender's parameters are the same as given in Table 7.5

through Table 7.10, while the attacker's parameters are defined by the following rules: Rule (1) an interval radius $\mu \geq 0$ is used, as shown in the bottom x axis of Fig. 7.11; Rule (2) all the monetary parameters (i.e., $\tilde{L}_y, C_a$) are bounded in the interval $[\sigma^{nominal} \cdot (1 - \mu), \sigma^{nominal} \cdot (1 + \mu)]$, and the $\sigma^{nominal}$ are the nominal values of the attackers' parameters as given in Table 7.5 through Table 7.10. In experiment s1, all the probabilistic parameters (i.e., $\tilde{P}_i^Z, \tilde{P}_i^P, \tilde{p}_y$) are bounded in the interval $[\sigma^{nominal} \cdot (1 - \mu), \sigma^{nominal} \cdot (1 + \mu)] \cap [0, 1]$, and the $\sigma^{nominal}$ are the nominal values of the attackers' parameters as given in Table 7.5 through Table 7.10. In experiment s2, all the probabilistic parameters (i.e., $\tilde{P}_i^Z, \tilde{P}_i^P, \tilde{p}_y$) are the same as the nominal values as given in Table 7.5 through Table 7.10.

Results shown in Fig. 7.11 demonstrate that the increase of interval radius μ would result in a decrease of the defender's optimal payoff. When $\mu = 0.0$, which means no interval uncertainty exists, the defender could have a payoff equal to the payoff from the BSE. When $\mu \geq \mu^*$ (in experiment s1, $\mu^* = 0.12$ for IBGS and $\mu^* = 0.16$ for ICGS, while in s2, $\mu^* = 0.46$ for IBGS and $\mu^* = 0.50$ for ICGS), the defender's payoff could be as low as her MaxiMin payoff. This means that, in the Bayesian Stackelberg CPP game, if the defender could not effectively bound the attacker's parameters into a relatively narrow interval, then her information of the attacker is useless.

Figure 7.11 also shows that with the same interval radius, the ICGS solution could always bring the defender a higher payoff than the IBGS solution, supporting our third conclusion.

7.2   Case Study #2: Applying the CCP Game for Scheduling Patrolling in the Setting of a Chemical Industrial Park

7.2.1   Case Study Setting

The layout of the cluster, the graph model, and the patrolling graph model of the case study are given in Figs. 6.1, 6.2, and 6.3 in Chap. 6. The total patrolling time T is set as 30 time slices. The patroller's driving time between different plants and patrolling time in each plant are shown in Table 6.2 in Chap. 6. Some more parameters and simplification assumptions of the case study are given hereafter.

For the sake of simplicity, we assume that the attacker has only one attack scenario and this scenario lasts for ten time slices in each plant. Table 7.34 gives the model inputs, i.e., the defender's reward (R^d) and loss (L^d) of detecting and not detecting an attacker; the attacker's gain (G^a) and penalty (P^a) from a successful and from a failed attack; the probability (f_{cpp}) that countermeasures in each plant can detect the attacker. The probability that the patroller can detect an attacker (i.e., σ_r, definition given in Fig. 6.4 in Chap. 6) should also be provided by security experts. However, for the sake of simplicity, we assume that in each time slice, if the attacker and the patroller stay in the same plant (i.e., an overlap situation), there is a

Table 7.34 Further model inputs for the case study of CCP game

	R^d	L^d	G^a	G^{a_min}	G^{a_max}	P^a	$f_{cpp} \& \tilde{f}_{cpp}$	$\tilde{f}_{cpp}^{min}$	$\tilde{f}_{cpp}^{max}$
'A'	1	16	10	9.5	10.2	3	0.45	0.44	0.46
'B'	1	11.2	6	5.5	6.4	3	0.3	0.29	0.31
'C'	1	14	8.3	8	8.5	3	0.42	0.41	0.43
'D'	1	12	7.1	7	7.4	3	0.45	0.44	0.46
'E'	1	15	10	9.5	10.3	3	0.5	0.49	0.51

probability of 0.05 that the attacker would be detected by the patroller. The unit of all the monetary parameters can be, for instance, k€.

It is worth noting that all these data concern estimations from the patroller. Therefore, the numbers of rewards (R^d), losses (L^d), and the detection probability (f_{cpp}) of countermeasures of each plant are the patroller's estimation of her own data. The amounts of the attacker's gains (G^a), penalties (P^a), and the attacker's estimation of the detection probability ($\tilde{f}_{cpp}$) of countermeasures of each plant, are the patroller's estimation of the attacker's data. For instance, "the gain of a successful attack on plant 'A' is 10" means that the patroller thinks the attacker will receive a value of 10 from this attack. The patroller may have uncertainties on guessing the attacker's parameters. Therefore, G^{a_min} and G^{a_max} are introduced to denote the patroller's minimal and maximal guesses of the attacker's gain of a successful attack. Similarly, $\tilde{f}_{cpp}^{min}$ and $\tilde{f}_{cpp}^{max}$ denote the patroller's minimal and maximal guesses of the attacker's estimation of the detection probability of countermeasures of every plant. The attacker's penalty of a failed attack is easier to estimate. For the simplicity of the model, we assume that the patroller can correctly guess the exact number of the penalty of a failed attack.

7.2.2 Game Modelling

There are two players in the case study game, namely the patroller and the attacker. Since only one attack scenario is considered, the attacker therefore has $m = 5 \times 30 \times 1 = 150$ pure strategies, being attack a plant (i.e., one of 'A', 'B', 'C', 'D', and 'E') at a time (i.e., at a time $t \in \{0, 1, 2, \ldots, 29\}$). The patroller has 435 possible actions that she can take, shown as edges in Fig. 6.3 (in Chap. 6) and therefore the patroller's strategy can be represented as a vector of 435 entries.

According to Formulas (6.13), (6.14), (6.20), and (6.21), the attacker and the patroller's payoffs can be calculated. Payoffs will be represented as linear polynomials of the patroller's strategy (i.e., $\vec{c}$), while the attacker's strategy determines the coefficients of the polynomials.

7.2.3   CCP Game Results

7.2.3.1   Stackelberg Equilibrium

Figure 7.12 shows the Modified Stackelberg Equilibrium (MSE) of the game developed for the case study, calculated by the MultiLPs algorithm shown in Table 6.5 and then slightly moved with a $\alpha = 0.1$. The black (and bold) lines demonstrate the patroller's optimal patrolling strategy. The associated number on the line denotes the probability that the defender will take this action. For instance, $c1 = 0.2275$ means that at time 0, the patroller should drive to node 'B2' at a probability of 0.2275. Furthermore, in patrolling practice, if the patroller arrives at a node in the figure, the conditional probabilities of the following actions can be calculated by Formula (6.6) in Chap. 6. For instance, the probability that the patroller would arrive at the red node $(6, \ 'C')$ in Fig. 7.12 is $sP_v = 0.4173$, and the conditional probabilities that the patroller should take the two actions (i.e., either patrolling in plant 'C' for a period of six time slices or driving to entrance 'B1' by a driving time of three time slices) are $cP_1 = \frac{0.2078}{0.4173} = 0.4979, cP_2 = \frac{0.2096}{0.4173} = 0.5021$ respectively.

The attacker's best response strategy in the MSE is to attack plant 'E' at time 9, shown in Fig. 7.12 as a red bold line. The short blue lines above the attacker's best response strategy line (i.e., the red bold line) represent the defender's patrolling actions which have a probability of being taken (i.e., $c > 0$) and would have overlap with the attacker's best response strategy. Table 7.35 shows the detail information of the defender's actions that have overlaps with the attacker's best response strategy. The 'Edge' column denotes the edge index (in Fig. 6.3 in Chap. 6) of the action. The c column shows the probability that the actions would be taken in the MSE, and these numbers are also shown in Fig. 7.12. The 'Overlap' column illustrates the time period that the actions overlap with the attacker's best response strategy. The 'σ' column provides the probability that the attacker would be detected by the corresponding action, and this probability is simply calculated as 0.05 multiplied by the overlapping time slices. For instance, edge 25 represents the patroller's action of patrolling plant 'E' from time 6 until time 13 while the attacker starts his attack in plant 'E' at time 9. Therefore, edge 25 overlaps with the attacker's attack in time zone [9,13], and the σ is $0.05 \times (13 - 9) = 0.20$.

Based on the results in Table 7.35, recalling Formula (6.11) and the τ_r calculation algorithm in Chap. 6, we have that:

$$f_p = \sum_r \tau_r \cdot \sigma_r = 0.0891$$
$$f = 1 - (1 - 0.5) * (1 - f_p) = 0.0949,$$
$$u_a = 2.88311 \ and \ u_d = -6.2407.$$

Table 7.36 further shows the information of all the defender's actions that have a probability of being played. The 'From' and 'To' columns denote the start and end node of the action, if shown in the patrolling graph.

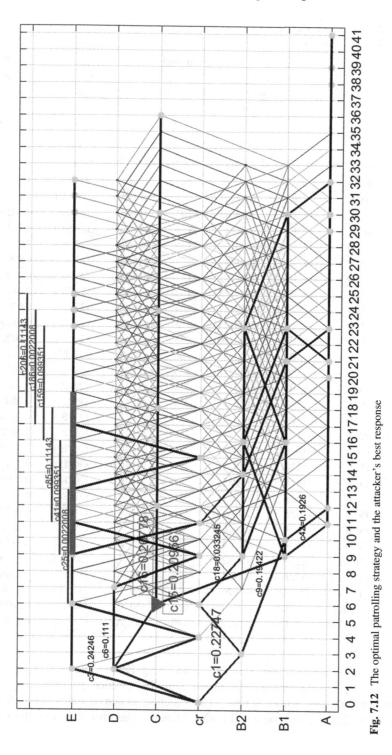

Fig. 7.12 The optimal patrolling strategy and the attacker's best response

Table 7.35 The patroller's actions that may detect the attacker

Edge	c	Overlap	σ
25	0.0022	[9,13]	0.20
41	0.0994	[9,16]	0.35
85	0.1114	[11,18]	0.35
159	0.0994	[16,19]	0.15
186	0.0022	[17,19]	0.10
206	0.1114	[18,19]	0.05

Table 7.36 The patroller's optimal strategy from the Stackelberg Equilibrium

Edge	From	To	c	Edge	From	To	c
1	(0,'cr')	(3,'B')	0.22747	98	(12,'A')	(21,'A')	0.19422
2	(0,'cr')	(2,'D')	0.53008	102	(13,'E')	(15,'cr')	0.002201
3	(0,'cr')	(2,'E')	0.24246	124	(14,'B')	(21,'B')	0.1422
4	(2,'D')	(4,'cr')	0.001299	155	(15,'cr')	(17,'E')	0.002201
5	(2,'D')	(6,'C')	0.41734	159	(16,'E')	(23,'E')	0.099351
6	(2,'D')	(7,'D')	0.11143	161	(16,'B')	(23,'B')	0.010434
8	(2,'E')	(9,'E')	0.24155	162	(16,'B')	(23,'B')	0.014701
9	(3,'B')	(10,'B')	0.19422	164	(16,'B')	(23,'B')	0.010409
11	(3,'B')	(6,'cr')	0.033245	165	(16,'B')	(23,'B')	0.014661
14	(4,'cr')	(6,'E')	0.002201	186	(17,'E')	(24,'E')	0.002201
15	(6,'C')	(9,'B')	0.20956	206	(18,'E')	(25,'E')	0.11143
16	(6,'C')	(12,'C')	0.20778	208	(18,'C')	(24,'C')	0.20778
18	(6,'cr')	(9,'B')	0.033245	238	(20,'A')	(29,'A')	0.1926
25	(6,'E')	(13,'E')	0.002201	258	(21,'A')	(30,'A')	0.050817
26	(7,'D')	(9,'cr')	0.11143	259	(21,'A')	(23,'B')	0.14341
40	(9,'E')	(11,'cr')	0.1422	260	(21,'B')	(23,'A')	0.1422
41	(9,'E')	(16,'E')	0.099351	299	(23,'E')	(30,'E')	0.099351
42	(9,'B')	(11,'A')	0.1926	301	(23,'B')	(30,'B')	0.16425
43	(9,'B')	(16,'B')	0.008503	304	(23,'B')	(30,'B')	0.029362
44	(9,'B')	(16,'B')	0.008457	316	(23,'A')	(32,'A')	0.1422
46	(9,'B')	(16,'B')	0.016631	328	(24,'E')	(31,'E')	0.002201
47	(9,'B')	(16,'B')	0.016614	330	(24,'C')	(30,'C')	0.20778
51	(9,'cr')	(11,'E')	0.11143	348	(25,'E')	(32,'E')	0.11143
52	(10,'B')	(12,'A')	0.19422	405	(29,'A')	(38,'A')	0.1926
79	(11,'cr')	(14,'B')	0.1422	415	(30,'A')	(39,'A')	0.050817
82	(11,'A')	(20,'A')	0.1926	417	(30,'B')	(32,'A')	0.19361
85	(11,'E')	(18,'E')	0.11143	420	(30,'C')	(36,'C')	0.20778
87	(12,'C')	(18,'C')	0.20778	430	(32,'A')	(41,'A')	0.33581

Let us now compare the Modified Stackelberg Equilibrium with the purely randomized patrolling strategy. In current patrolling practice, patrollers may randomly schedule their patrolling route. This situation, as demonstrated in Fig. 6.3 in Chap. 6, is simply assigning equal probabilities to edges that start from the same

Table 7.37 Comparison of the CCP MSE strategy and the purely randomized strategy

Edge	Overlap	c	rc	σ
82	[11,19]	0.1926	0.0046	0.4
98	[12,19]	0.1942	0.0139	0.35
156	[15,19]	0	0.0019	0.2
176	[16,19]	0	0.0071	0.15
196	[17,19]	0	0.0024	0.1
216	[18,19]	0	0.0039	0.05
425	[9,10]	0	0.0100	0.05
430	[9,11]	0.3358	0.0274	0.1

node. For instance, at the starting node (i.e., $(0, \acute{c}r)$), the patroller would come to plant (entrance) 'B2', 'D', and 'E' with the same probability, being 1/3.

In the case study, if the defender would purely randomize her patrolling, then the attacker's best response would be attacking plant 'A' at time 9. The attacker and the defender would obtain a payoff of 4.0653 and −8.2393, respectively. Compared to the Modified Stackelberg Equilibrium of the CCP game, the defender's payoff reduces from −6.2407 to −8.2393.

Table 7.37 illustrates the differences between the CCP MSE strategy and the purely randomized strategy. The edge column shows the edges in the patrolling graph showing an overlap with the attacker's best response strategy to the defender's purely randomized strategy (i.e., attack plant 'A' at time 9). The overlap column shows the period of the attack procedure being overlapped by the edge. The 'c' and 'rc' columns show the probability that the patroller will follow the edge, resulting from the CCP MSE strategy and from the purely randomized strategy, respectively. The 'σ' column shows the probability that the attacker will be detected by the patroller by the action she undertakes, represented by this edge.

With the results in Table 7.37, the probability that the attacker would be detected can be calculated, being $f_p^c = 0.1786$ and $f_p^{rc} = 0.0118$, for the defender's CCP MSE strategy and for the defender's purely randomized strategy, respectively. This result reveals that the CCP MSE strategy is characterized with a higher probability that the attacker is detected at plant 'A', and thus enforces the attacker to attack plant 'E' instead of attacking plant 'A'.

Furthermore, in current patrolling practice, some patrollers may follow a fixed patrolling route. In the patrolling graph, if we further constraint the probability that an action (an edge) is taken to be either 0 or 1, that is, $c \in \{0, 1\}$ instead of $c \in [0, 1]$, then a vector of c that satisfies Formulas (6.4) and (6.5) from Chap. 6, represents a fixed patrolling route. The bold route shown in Fig. 7.13 is the optimal fixed patrolling route considering intelligent attackers. The route is that: the patroller starts from 'cr'; she goes to plant 'D' and patrols plant 'D'; after then, she goes to plant 'A' and patrols 'A'; she further goes to entrance 'B1' and then comes back to plant 'A' and patrols plant 'A'. The red dot line in Fig. 7.13 denotes the attacker's best response strategy to the optimal fixed patrolling route, and it is, attacking plant 'C' at time 21. If the defender follows the fixed patrolling route and the attacker plays his

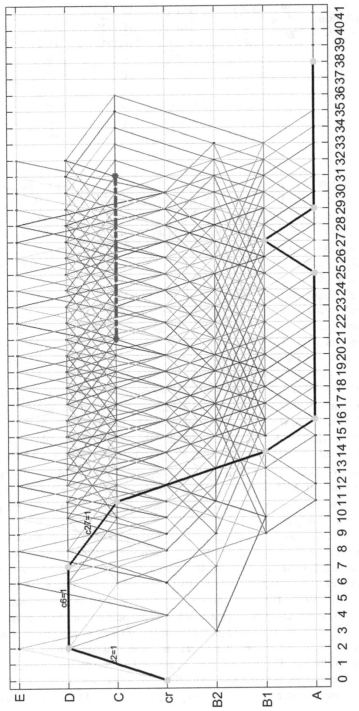

Fig. 7.13   The patroller's optimal fixed patrolling route and the attacker's best response

best response, as shown in Fig. 7.13, the payoffs for the defender and for the attacker
are -7.7 and 3.5540 respectively.

It is worth noting the defender's optimal fixed patrolling route is not unique and
the attacker's best response is not unique as well. For instance, knowing the
patroller's fixed route, the attacker would be indifferent by starting his attack at
any time. However, the defender and the attacker's payoff would not be different.
Therefore, here we only show one optimal fixed patrolling route and one attacker's
best response strategy.

7.2.3.2 Robust Equilibrium

Figure 7.14 shows the robust solution of the Interval Chemical Cluster Patrolling
game, based on the input data from Table 7.34. Table 7.38 shows all the edges
having a probability higher than zero. Notations of Fig. 7.14 and Table 7.38 are the
same as defined in Fig. 7.12 and Table 7.36. The attacker's strategy of attacking
plant 'E' at time 0 has the highest lower bound payoff, shown as a red bold line in
Fig. 7.14. Furthermore we have:

$$f_p = 0.10805 \cdot 0.35 + 0.06043 \cdot 0.05 + 0.00751 \cdot 0.05 + 0.03415 \cdot 0.10 = 0.0446$$
$$f = 1 - \left(1 - \tilde{f}_{cpp}^{\,max}\right) \cdot \left(1 - f_p\right) = 0.5319$$
$$R = G^{a\_min} \cdot \left(1 - f\right) - P^a \cdot f = 2.8516$$

Figure 7.15 shows the attacker's payoff information of the robust solution of the
Interval CCP game. As also demonstrated in the figure, different sub-figures denote
the attacker's payoff by attacking different plants. The x-axis denotes the start time
of attacks and therefore a combination of an x coordinate and a certain sub-figure
represents an attacker strategy. The vertical lines denote the range of the patroller's
estimation of the attacker's payoffs, under the conditions that the patroller plays her
strategy shown in Table 7.38 and the attacker plays the corresponding strategy (i.e.,
the sub-figure and the x coordinate). Horizontal lines in all sub-figures have the same
y value, and it is the attacker's highest lower bound payoff (i.e., R). A red square dot
means that the corresponding attacker strategy is the attacker's possible best
response strategy while a green circle dot means that the corresponding strategy is
not a possible best response strategy for the attacker.

As shown in Fig. 7.15, for an attacker strategy, if the attack target is not plant 'E'
and, if the strategy has an upper bound payoff higher than R, then the attacker
strategy is thought to be a possible best response for the attacker (i.e., a red square is
used), otherwise if the strategy has an upper bound payoff lower than R, then it is
considered not to be a possible best response (i.e., a green dot is used). If an attacker
strategy aims to attack plant 'E', then the above rule does not work, as shown in
sub-figure 'Plant E'. The reason is that, the robust solution is achieved when the
attacker plays a strategy of attacking plant 'E' at time 0. Therefore, whether
strategies which aim at attacking plant 'E' should be possible best response strategies

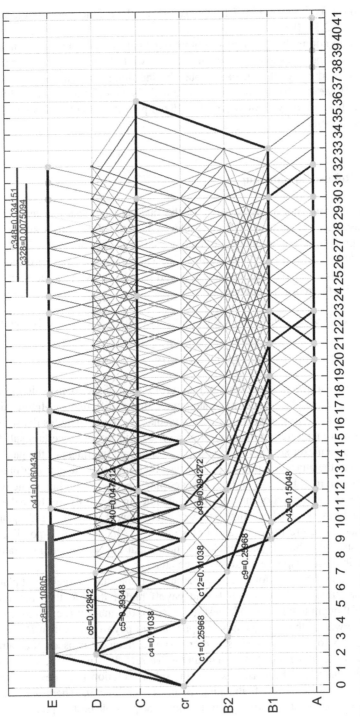

Fig. 7.14 Robust solution of the interval CCP game

Table 7.38 The patroller's optimal strategy from the robust solution

Edge	From	To	c	Edge	From	To	c
1	(0,'cr')	(3,'B')	0.25968	104	(13,'D')	(15,'cr')	0.007509
2	(0,'cr')	(2,'D')	0.63228	121	(14,'B')	(21,'B')	0.11038
3	(0,'cr')	(2,'E')	0.10805	124	(14,'B')	(21,'B')	0.040103
4	(2,'D')	(4,'cr')	0.11038	155	(15,'cr')	(17,'E')	0.007509
5	(2,'D')	(6,'C')	0.39348	159	(16,'E')	(23,'E')	0.060434
6	(2,'D')	(7,'D')	0.12842	186	(17,'E')	(24,'E')	0.007509
8	(2,'E')	(9,'E')	0.10805	206	(18,'E')	(25,'E')	0.034151
9	(3,'B')	(10,'B')	0.25968	208	(18,'C')	(24,'C')	0.24299
12	(4,'cr')	(7,'B')	0.11038	219	(19,'B')	(26,'B')	0.094272
15	(6,'C')	(9,'B')	0.15048	238	(20,'A')	(29,'A')	0.15048
16	(6,'C')	(12,'C')	0.24299	258	(21,'A')	(30,'A')	0.094272
26	(7,'D')	(9,'cr')	0.12842	259	(21,'A')	(23,'B')	0.16519
29	(7,'B')	(14,'B')	0.11038	260	(21,'B')	(23,'A')	0.15048
40	(9,'E')	(11,'cr')	0.047612	299	(23,'E')	(30,'E')	0.060434
41	(9,'E')	(16,'E')	0.060434	301	(23,'B')	(30,'B')	0.16519
42	(9,'B')	(11,'A')	0.15048	316	(23,'A')	(32,'A')	0.15048
49	(9,'cr')	(12,'B')	0.094272	328	(24,'E')	(31,'E')	0.007509
51	(9,'cr')	(11,'E')	0.034151	330	(24,'C')	(30,'C')	0.24299
52	(10,'B')	(12,'A')	0.25946	348	(25,'E')	(32,'E')	0.034151
79	(11,'cr')	(14,'B')	0.040103	361	(26,'B')	(33,'B')	0.094272
80	(11,'cr')	(13,'D')	0.007509	405	(29,'A')	(38,'A')	0.15048
82	(11,'A')	(20,'A')	0.15048	415	(30,'A')	(39,'A')	0.094272
85	(11,'E')	(18,'E')	0.034151	417	(30,'B')	(32,'A')	0.16519
87	(12,'C')	(18,'C')	0.24299	420	(30,'C')	(36,'C')	0.24299
95	(12,'B')	(19,'B')	0.094272	430	(32,'A')	(41,'A')	0.31567
98	(12,'A')	(21,'A')	0.25946	435	(33,'B')	(36,'C')	0.094272

will be determined by constraint c5 in Formula (6.22) in Chap. 6, instead of by the payoff range constraint (i.e., Constraint c4 in Formula (6.22)).

7.3 Conclusion

In this chapter, two case studies are defined and investigated, for illustrating the Chemical Plant Protection (CPP) game and for illustrating the Chemical Cluster Patrolling (CCP) game respectively.

Results of case study #1 reveal that in the CPP game, the defender's uncertainties about the attacker's information reduce the defender's expected payoff. In case that the defender has deep uncertainties about the attacker, for instance, both on the attacker's parameters and on the attacker's rationality, the defender's expected payoff from robust solutions (e.g., the interval CPP game solution or the MoSICP

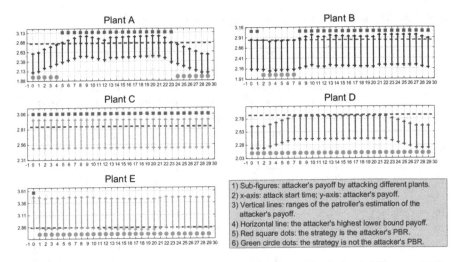

Fig. 7.15 Attacker payoff information of the robust solution of the Interval CCP game (PBR: possible best response)

solution) would not be much higher than her payoff from the MiniMax solution. Therefore, due to the lack of security data in current industrial practice, managers' security efforts in chemical plants tend to try to reduce (minimize) the consequences of the worst (maximal) scenarios they consider.

Results of case study #2 show that by strategically randomizing patrolling routes, the patroller would have higher expected payoffs, indicating that patrolling more hazardous plants would be more likely (that is, they are accompanied by higher probabilities for the patroller). The performance of the patrolling strategy from the Stackelberg equilibrium is highly better than the performance of the purely random-ized patrolling routes and the performance of any fixed patrolling route.

References

1. API. Security risk assessment methodology for the petroleum and petrochemical industries. In: 780 ARP, editor. 2013.
2. Lee Y, Kim J, Kim J, Kim J, Moon I. Development of a risk assessment program for chemical terrorism. Korean J Chem Eng. 2010;27(2):399–408.
3. Reniers G, Cozzani V. Domino effects in the process industries: modelling, prevention and managing. Amsterdam: Elsevier B.V.; 2013. p. 1–372.
4. Zhuang J, Bier VM. Balancing terrorism and natural disasters-defensive strategy with endoge-nous attacker effort. Oper Res. 2007;55(5):976–91.
5. Lemke CE, Howson J, Joseph T. Equilibrium points of bimatrix games. J Soc Ind Appl Math. 1964;12(2):413–23.
6. Conitzer V, Sandholm T, editors. Computing the optimal strategy to commit to. In: Proceedings of the 7th ACM conference on electronic commerce. ACM; 2006.

7. Jiang AX, Nguyen TH, Tambe M, Procaccia AD, editors. Monotonic maximin: a robust stackelberg solution against boundedly rational followers. In: International conference on decision and game theory for security. Springer; 2013.
8. Nguyen TH, Jiang AX, Tambe M, editors. Stop the compartmentalization: unified robust algorithms for handling uncertainties in security games. In: Proceedings of the 2014 international conference on autonomous agents and multi-agent systems. International Foundation for Autonomous Agents and Multiagent Systems; 2014.

Chapter 8
Conclusions and Recommendations

Chemicals-using industries have an important role in modern society for providing the basic ingredients (fuels, chemicals, intermediates and consumer products) for our modern day lives and luxury. However, they also pose huge threats to society due to the mere use and storage of large amounts of hazardous materials with sometimes extreme processing conditions. The prevention of unintentionally caused events, which is the field of occupational safety and process safety, has been significantly improved in the process industries. Conversely, the physical protection of chemical plants and areas from malicious attacks, being the field of physical security, has not received enough attention yet by both academic researchers and industrial practitioners.

Several qualitative and semi-quantitative security risk assessment methods have been published. For instance, the Security Risk Factor Table (SRFT) and the American Petroleum Institute recommended standard on "Security Risk Assessment Methodology for Petroleum and Petrochemical Industries" (the API SRA). These conventional security risk assessment methods, though been currently used in industrial practice in the United States, have the drawback that they are not able to consider intelligent interactions between the defender and the potential attackers.

To counter the current disadvantage of security risk assessments, we introduce game theory as a decision-support mathematical approach for managing security in chemical industrial areas. The Chemical Plant Protection (CPP) game, which purpose it is to optimally set security alert levels at every entrance and every zone in chemical plants, and the Chemical Cluster Patrolling (CCP) game, which can be employed to randomly but strategically schedule security guard patrolling among different plants, are elaborated. Extensions are also proposed to deal with the defender's uncertainties on attacker parameters, both for the CPP game and for the CCP game.

Eight conclusions are formulated. Nine recommendations are given based on the conclusions that we draw.

© Springer International Publishing AG, part of Springer Nature 2018
L. Zhang, G. Reniers, *Game Theory for Managing Security in Chemical Industrial Areas*, Advanced Sciences and Technologies for Security Applications, https://doi.org/10.1007/978-3-319-92618-6_8

Conclusion 1

Conventional security assessment methods, such as the SRFT and the API SRA, are mostly developed by senior security experts with plenty of experience and expertise on physical security policies and management. Therefore, these methods have the advantage of being practically implementable in industrial practice and of being understandable and used by practitioners. However, these methods are mainly qualitative or semi-quantitative, and are not able to provide adequate information for decision makers to quantitatively allocate security resources; we refer for this observation to arguments given in Cox [1, 2]. Furthermore, failing to model the intelligent interactions between defender and attackers, these conventional methods may lead to incorrect results with respect to the allocation of security resources; see also further arguments in Powell [3].

Game theory, conversely, developed by mathematicians and economists, is quite abstract to the practice of chemical security management. However, game theory has the advantage on modelling strategic decision making in a multiple players setting and on providing quantitative results as output information. Several game theory-based security systems have been developed and implemented, such as the ARMOR system for the Los Angeles airport, the PROTECT system for the US coast guard, and the IRIS system for the Federal Air Marshal's service etc. [4]. In this thesis, we developed such an analogous game-theory-based security system for the chemical industry.

- **Recommendation 1**
  To improve security within the chemical industry, conventional security risk assessment methods and game theory need to be integrated. In the integrated framework, game theoretical models need to be provided with inputs from conventional methods, and game theoretical results need to be 'translated' to industrial practice.

Conclusion 2

There are many types of security countermeasures. Even in the situation of a limited budget, the defender can still combine several countermeasures, in order to secure her assets. In conventional security risk assessment methods, the effectiveness of a bundle of countermeasures is not assessed. For instance, in the API SRA methodology, the SRA team only re-estimates vulnerabilities and consequences presuming that one proposed countermeasure is implemented (see Form 6 in the API SRA document [5]). However, synergistic effects of multiple countermeasures should not be under-estimated. An example of a synergistic effect is the combination of a camera system and having fences. Cameras without fences or fences without cameras are much less efficient than both together.

- **Recommendation 2**
  Risks reductions by bundles of countermeasures should also be estimated. In fact, there might be a great number of such bundles. For instance, in case of a recommendation of in total 10 countermeasures and bundles existing of

two countermeasures, there can be $2^{10} = 1024$ of these bundles of counter-measures. However, the number can be significantly reduced by budget constraints as well as by using field knowledge.

Conclusion 3

As already mentioned, an important challenge of assessing and managing security risks in chemical plants is that the defender deals with intelligent adaptive adversaries. To fight with these intelligent attackers, the defender should not only pay attention to her own interests, but also study the attacker's interests, since intelligent attackers may exhibit a high probability to attack a target which from a safety viewpoint is quite safe.

Conventional security risk assessment methods, however, mainly focus on the defender's interests and implicitly assume that the attacker has opposite interests to the defender. Nonetheless, potential attackers within the chemical industry are various and different attackers have different goals. Therefore, it is not necessary that attackers always have an opposite interest to the defender.

- **Recommendation 3**

  In a security risk assessment procedure, attention should be paid to the data assessment from the adversaries' viewpoint. Putting "the defender's feet also in the attackers' shoes" can be helpful for gaining insights by company security management.

Conclusion 4

In a deep uncertain case, which means that the defender has huge uncertainties on the attackers' interests, rationalities, and capabilities etc., the defender is better off if she minimizes her worst/maximal loss. Therefore, although we conclude in *conclusion 3* that the defender should pay attention to learn the attackers' interest, in the current stage, if the learning is too difficult (e.g., due to the lack of reliable data), ignoring the attacker's interests can be a feasible solution for the defender.

- **Recommendation 4**

  Even from a game theoretic point of view, conventional security risk assessment methods have their rationales on implicitly assuming that the attackers have opposite interests to the defender's interest. Due to the difficulties of obtaining knowledge and data about the attackers, huge uncertainties of the attackers may exist. In this case, the defender is secure to play her MiniMax solution, which is also the optimal solution in a zero-sum game.

Conclusion 5

Risk scoring methods are still extensively used in security risk assessment procedures, after being proved theoretically incorrect [6]. Moreover, on the one hand practitioners say it is difficult to obtain quantitative data, while on the other hand in some qualitative methods, the security risk management team decides a security risk score based on quantitative descriptions. An example of such practice can be found as Table 3.4 in Chap. 3 of this book.

- **Recommendation 5**

  The security risk assessment team should work on quantitative data directly, instead of transferring these data into scores. Quantitative data extracted from industrial practice are often associated with uncertainties, for instance, instead of knowing an exact number of the consequence of a certain event, it is more likely that we know a lower and an upper bound of the value of the consequence. Current game theoretical models are able to deal with data with this type of uncertainties, (see for instance Chap. 4 of this book) and should therefore be used in security risk assessment.

Conclusion 6

In case that the attackers' interests are not strictly opposite to the defender's interests, which means that the security game is not a strategic zero-sum game, then the defender's payoff from a sequential game is higher and more stable than her payoff from a simultaneous game. Otherwise, if the defender believes that the attackers always have strictly opposite interests to her, then it does not matter whether the game is played sequentially or simultaneously.

- **Recommendation 6**

  In a situation that the security information of a chemical plant is publicly known (thus the common knowledge assumption of a game can easily hold), then for defending those premediated attackers (premediated attackers are more likely to be strategic attackers, e.g., an IS terrorist), industrial managers are suggested to make their security plan public, to deter and stop those attackers.

Conclusion 7

Being mathematically complicated and being too abstract for industrial practice prevent game theory to be more popular among industrial practitioners. As we may notice from Chaps 3, 4, 5, 6, and 7 in this book, game theory uses plenty of mathematical formulas and numbers, and regretfully, at least for optimal decision-making support, chemical security related terminologies (e.g., assessing vulnerabilities, threats, etc.) does not. Industrial practitioners doubt the usefulness of these formulas and the practical meanings of these numbers.

Furthermore, the correctness of results from game theoretic models strictly relies on the assumptions that the modeller uses. Some assumptions used in game theoretic models are quite unrealistic, e.g., the 'common knowledge' assumption. Therefore, industrial practitioners doubt the correctness of game theoretic results.

- **Recommendation 7.1**

  User-friendly interfaces should be developed for game theoretic models. With the interface, a security risk assessment team can use game theoretic models as a black-box tool, and this way, it is possible for security managers

Fig. 8.1 An extended framework of integrating conventional security risk assessment methods and security game

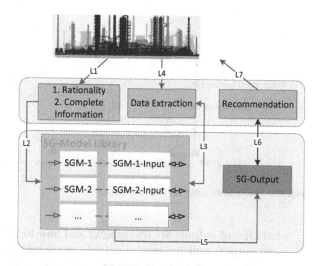

to only have to provide the black-box tool with input data and afterwards to analyse the outputs of the tool.

Figure 8.1 (an extension of Fig. 2.3 in Chap. 2) shows an extended framework of integrating conventional security risk assessment methods and security game theory. In the first step (L1), the security risk assessment team should evaluate what kind of threats the plant is faced with. Moreover, based on the current information and the team's judgements, the team should estimate whether these potential attackers are rational players or not, and they should estimate how much information the team has about the attackers. In the second step (L2), the team chooses a proper security game model from the so-called "security game model library" and learns what kind of input data is needed for the chosen security game model. In the third step (L3, L4), the team extracts the needed input data, by using a conventional security risk assessment method, the API SRA, for instance. In the fourth step (L5), the team simply runs the chosen security game model without necessarily knowing the details of the model. In the fifth step (L6, L7), the team translates the outputs of the chosen security game model into implementable recommendations.

In Fig. 8.1, steps L1, L2, L4, and L7 are closely related to the practice of industrial security, and therefore they can be carried out by a security risk assessment team independently. Steps L3 and L6 should be done cooperatively by an SRA team and a security game developer. In step L3, the game developer informs the SRA team what kind of data is needed and what are the meanings of the data. In the meantime, the SRA team judges whether the data is achievable. If the answer is 'yes', then the game developer and the SRA team discuss the data structure of the inputs, while if the answer is 'no', then the game developer must revise the security game to be able to deal with achievable data. In step L6, the SRA team and the game developer discuss

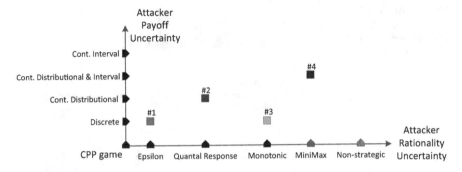

Fig. 8.2 Uncertainty space for the CPP game

what kind of outputs are meaningful and how to build the map between the game outputs and the implementable recommendations. Step L5 concerns purely game theoretic calculations, and the SRA team should not pay attention to this step.

In summary, the bottom grey part of Fig. 8.1 should be a black-box for the SRA team.

- **Recommendation 7.2**

  Game theoretic models for dealing with various uncertainties should be developed. In other words, the SG-Model Library in Fig. 8.1 should be complete, to make sure that whatever the result of 'L2' is, a security game model always exists.

  Fortunately, developments on computational game theory have provided models and algorithms for studying games played by bounded rational players and games where 'common knowledge' does not hold. Figure 8.2 (adopted from Zhang and Reniers [7]) shows the uncertainty space of the Chemical Plant Protection game (CPP game) [8]. The origin point is the CPP game with rational players and common knowledge assumptions. The x-axis represents the attacker's rationality, such as the epsilon-optimal attackers, quantal response attackers, etc. The y-axis denotes the defender's uncertainty on the attacker's payoffs, such as the discrete uncertainty, Bayesian uncertainty, interval uncertainty, etc. Each point in the uncertainty space corresponds to a realistic situation and a cluster of models and algorithms. If the uncertain space of the Chemical Cluster Patrolling (CCP) game would be plotted, a third dimension named "uncertainty on the attacker's observation" should also be added.

  The output of 'L2' in Fig. 8.1 decides a coordinate in Fig. 8.2. Therefore, models and algorithms should be developed for all the meaningful coordinates in Fig. 8.2. To achieve this goal, models and algorithms for dealing with combinations of multiple types of uncertainties need to be enhanced. There are abundant studies on dealing with a single type of uncertainty, i.e., points on axis in Fig. 8.2. However, in reality, a defender often faces multiple types of uncertainties, e.g., point #1 in Fig. 8.2 represents multiple types of attackers and each type of attackers are epsilon-optimal players [9].

Conclusion 8

A purely randomized patrolling route or a fixed patrolling route does not make best use of the security guard patrolling team. A purely randomized patrolling route fails to cover more hazardous plants more frequently. The downside of a fixed patrolling route is that the patroller's position may be predictable to an attacker. Game theory can therefore be used to generate random (thus being unpredictable) but strategic (thus patrolling higher hazardous plants more often) patrolling routes.

- **Recommendation 8**
   Security patrolling in current industrial practice should be re-thought and re-conceptualized by using game-theoretical models.

References

1. Cox LAT Jr. Some limitations of "Risk = Threat × Vulnerability × Consequence" for risk analysis of terrorist attacks. Risk Anal. 2008;28(6):1749–61.
2. Cox L. What's wrong with risk matrices? Risk Anal. 2008;28(2):497–512.
3. Powell R. Defending against terrorist attacks with limited resources. Am Polit Sci Rev. 2007;101 (03):527–41.
4. Tambe M. Security and game theory: algorithms, deployed systems, lessons learned. Cambridge: Cambridge University Press; 2011.
5. API. Security risk assessment methodology for the petroleum and petrochemical industries. In: 780 ARP, editor. 2013.
6. Cox LAT, Babayev D, Huber W. Some limitations of qualitative risk rating systems. Risk Anal. 2005;25(3):651–62.
7. Zhang L, Reniers G. Applying game theory for adversarial risk analysis in process plants. In: Reniers G, Khakzad N, van Gelder P, editors. Security risk assessment and management in the chemical and process industry. Berlin: De Gruyter; 2018.
8. Zhang L, Reniers G. A game-theoretical model to improve process plant protection from terrorist attacks. Risk Anal. 2016;36(12):2285–97.
9. Pita J, Jain M, Tambe M, Ordóñez F, Kraus S. Robust solutions to Stackelberg games: addressing bounded rationality and limited observations in human cognition. Artif Intell. 2010;174 (15):1142–71.

Printed in the United States
by Bookmasters

Printed in the United States
By Bookmasters